Commanding an Air Force Squadron

Timothy T. Timmons, Col, USAF

Research Fellow
National Defense University

AF677362

Air University Press
Maxwell Air Force Base, Alabama 36112-6428
in Coordination with
National Defense University Press
Fort Lesley J. McNair
Washington, DC 20319-6000

December 1993

Library of Congress Cataloging-in-Publication Data

Timmons, Timothy T., 1946–
Commanding an Air Force Squadron / Timothy T. Timmons
p. cm.
"December 1993."
Includes bibliographical references and index.
1. United States. Air Force—Officer's handbooks. 2. Command of troops. I. Title.
UG633.T53 1993 95-38707
358.4'133'0973—dc20 CIP

ISBN 1-58566-008-6

First Printing December 1993
Second Printing January 1995
Third Printing August 1995
Fourth Printing August 1997
Fifth Printing March 1999
Sixth Printing July 1999
Seventh Printing December 2000

Disclaimer

This publication was produced in the Department of Defense school environment in the interest of academic freedom and the advancement of national defense-related concepts. The views expressed in this publication are those of the author and do not reflect the official policy or position of the Department of Defense or the United States government.

This publication has been reviewed by security and policy review authorities and is cleared for public release.

To My Wife Linda,
My Partner in Life,
Whose Love and Devotion
For Nearly Two Decades
Have Inspired Me and
Allowed Me to Pursue My Dreams

Contents

Chapter *Page*

Chapter *Page*

Foreword

The privilege of commanding an Air Force squadron, despite its heavy responsibilities and unrelenting challenges, represents for many Air Force officers the high point of their careers. It is service as a squadron commander that accords true command authority for the first time. The authority, used consistently and wisely, provides a foundation for command. As with the officer's commission itself, command authority is granted to those who have earned it, both by performance and a revealed capacity for the demands of total responsibility. But once granted, it must be revalidated every day. So as you assume squadron command, bringing your years of experience and proven record to join with this new authority, you still might need a little practical help to succeed with the tasks of command.

This book offers such help. As a squadron commander, your duties and responsibilities will change your life. You will immerse yourself totally in the business of your squadron, the problems of its people, and the challenges of its mission. Nowhere else, at no other level of command, do those three elements blend so intimately, and at no other level is the mission so directly achieved. Here is where hydraulic fluid spills, where operations succeed or fail, where unit leadership is paramount, where the commander becomes very visible to senior Air Force leaders—who depend upon each commander to achieve his or her part of the mission and judge each accordingly.

Commanding an Air Force Squadron brings unique and welcome material to a subject other books have addressed. It is rich in practical, useful, down-to-earth advice from officers who have recently experienced squadron command. The author does *not* quote regulations, parrot doctrine, or paraphrase the abstractions that lace the pages of so many books about leadership. Nor does he puff throughout the manuscript about how *he* did it. Rather, he presents a digest of practical wisdom based on real-world experience drawn from the reflections of many former commanders from many different types of units. He addresses all Air Force squadron commanders, rated and nonrated, in all sorts of missions worldwide.

Colonel Timmons provides a useful tip or a shared experience for every commander and (very important) for every officer about to assume command. Witness, for example, how he begins—with a chapter about how to *take* command. In that chapter, you'll read about what to do, what not to do, and what to expect of others during that vital transition. It is a chapter that should be required reading for any officer selected for squadron command. Throughout the book, Colonel Timmons provides "Proverbs for Command," collections of capsule wisdom—what some might call "one-liners with punch." These proverbs, which have as much depth as brevity, are ones you will probably want to reread frequently as you focus on solving problems of the moment and heading off other problems before they develop.

Obviously directed at Air Force readers, Colonel Timmons's book—because of its aphoristic, anecdotal, concrete approach—will speak to readers in other services and in many civilian organizations and institutions as well. Shelves in the nation's bookstores today are groaning under the weight of "how-to" leadership books purporting to reveal the secrets of how to succeed in one career or another. Most do not last. Only a few stand the test of time; this promises to be one of them.

As the Air Force increases its vital role in the new world of joint operations, this book will become increasingly valuable for several reasons. It reveals to readers in other services those matters peculiar to commanding an Air Force squadron. It also shows that much of the "stuff" of command is common to all services. Both those revelations will help advance interservice unity in the new joint world. In the world of combined operations, wherein US forces serve with forces of other nations, this book also has a useful role. It shows in a realistic way the basic leadership concerns of Air Force squadron commanders. By so doing, it serves as an explanatory text to allied officers, as a model for leadership studies, and as a conduit through which officers of other nations may discover how remarkably similar their problems are to ours.

A great strength of this book lies in its style.

Written in plain English, it is an easily read and digested text. Rich in specific advice, enjoyable anecdotes, and collected proverbs of command, it will also serve as a splendid reference, standing ready to aid as you face the ever-shifting pressures of command responsibility. Colonel Timmons has said that he will consider his work successful if each reader finds just one useful bit of advice helpful to him or her as a squadron commander. Based on that criterion, *Commanding an Air Force Squadron* cannot miss.

One final observation is in order. General (and later Secretary of State) George C. Marshall, an architect of the Allied victory in World War II and author of the plan that led to European recovery after the war, once noted how much can be done if no one worries about who gets the credit. The author researched this book while he was a student at the National War College and a Research Fellow at National Defense University. Many of his NWC classmates of all services as well as members of the Industrial College of the Armed Forces contributed willingly, as did numerous other former squadron commanders, who shared their experiences in interviews and letters from around the globe. Colonel Timmons then continued work on the book after graduation, devoting a good deal of his off-duty time to the effort. The director and editors at the National Defense University Press helped refine the manuscript and prepare it for publication. When lack of resources prevented NDU Press from continuing, the Air University Press stepped in to publish this fine edition. All these persons and institutions contributed to the single, nonparochial goal of assuring that the manuscript would reach the audience for which it was intended, an audience that will greatly benefit for many years to come. General Marshall would be pleased. And I am sure readers will be as well.

B. C. HOSMER, Lt Gen, USAF
Superintendent
United States Air Force Academy

About the Author

Col Timothy T. Timmons

Col Timothy T. Timmons is chief of the Joint Training Branch, Joint Exercise and Training Division, Directorate for Operational Plans and Interoperability (J-7) on the Joint Staff. In this capacity, he is responsible for all policy issues pertaining to the chairman of the Joint Chiefs of Staff's (CJCS) joint training program.

Born in Cincinnati, Ohio, on 13 October 1946, he holds a bachelor of science degree in mathematics from Florida State University and a master of arts degree, also in mathematics, from the University of Cincinnati. He is a resident graduate of the National War College, the Armed Forces Staff College, and the Air Force Squadron Officer School and has also completed the Air War College, Air Command and Staff College, and Squadron Officer School by correspondence.

Colonel Timmons earned his navigator wings in January 1974 as a distinguished graduate from Undergraduate Navigation Training Class 74-12 and was initially assigned to the 60th Military Airlift Wing, Travis AFB, California, as a C-5A airlift navigator. During his tour at Travis he served as an instructor and flight examiner navigator, a flight simulator instructor, and an airlift control element commander. In April 1975 he crewed the first trans-Pacific, nonstop C-5 airlift mission from Travis to Tan Son Nhut AB, South Vietnam.

In January 1979 Colonel Timmons was reassigned to Headquarters, United States Air Force at the Pentagon. He served on the Air Staff for four years as a personnel analyst in the Directorate of Personnel Plans where he was heavily involved in analytical work supporting flight pay and other compensation initiatives. He was also the 1981 national chairman of the Air Force Association's Junior Officer Advisory Council.

From February to July 1983, Colonel Timmons attended Armed Forces Staff College in Norfolk, Virginia. In July 1983 he went overseas for a one-year remote assignment with the United Nations Truce Supervision Organization (UNTSO) in the Middle East. While there, he served six months in south Lebanon, performing unarmed patrols and week-long observation duty for Observer Group Lebanon and six months in Cairo, Egypt, as the air liaison officer for Observer Group Egypt.

In July 1984 Colonel Timmons was assigned to the 323d Flying Training Wing, Mather AFB, California. He completed Instructor Training School as a distinguished graduate and subsequently served as the chief of the Wing Scheduling Branch, operations officer of the 450th Flying Training Squadron (FTS), and chief of the Wing Standardization and Evaluation Division before being named commander of the 450th FTS on 6 March 1987. During his tenure as squadron commander, the 450th was first responsible for the Fighter, Attack, Reconnaissance track of Specialized Undergraduate Navigator Training (SUNT) and then for the initial 70-day "Core" phase of SUNT. This mission change resulted in the 450th becoming the largest flying training squadron in Air Training Command. On 15 March 1989, Colonel Timmons relinquished his command and was named as Mather's deputy base commander.

Colonel Timmons was selected as a research fellow at the National Defense University and a student at the National War College for the 1989-90 academic year. He finished his studies, which concentrated on Middle East/Gulf regional issues, in June 1990 and was then assigned to the Joint Staff. His initial duty there was as the exercise plans officer for USCENTCOM, USSOCOM, and FORSCOM exercises. Between August 1990 and March 1991, Colonel Timmons served as a member of the Chairman's Crisis Action Team in the National Military Command Center for Operations Desert Shield and Desert Storm. He assumed his present duties in July 1991.

Colonel Timmons' decorations include the Defense Meritorious Service Medal, the Meritorious Service Medal with two oak leaf clusters, the Joint Service Commendation Medal, the Combat Readiness Medal, the Armed Forces Expeditionary Medal, and the United Nations Medal.

Colonel Timmons is married to the former Linda Joan Shriver of Cincinnati, and they have two daughters, Michelle, age fifteen, and Nicole, age twelve.

Preface

Webster's Ninth New Collegiate Dictionary defines command as (1) to have or to exercise direct authority, (2) to give orders, (3) to be commander, (4) to dominate as if from an elevated place. However, as any military member can readily attest, command in the context of the armed forces is much, much more than any dictionary could ever describe. Military command is unique.

The same dictionary defines squadron as a unit of the US Air Force higher than a flight and lower than a group. Yet again, anyone who has ever served in an Air Force squadron will tell you that it is much more than the definition listed above—a squadron is a unique type of military unit.

Putting the two words together, SQUADRON COMMAND has special meaning to the Air Force because it is normally the first time in an officer's career that he or she gets the opportunity to exercise command authority in the true sense of the word. Thus, this book brings two particular military concepts together at a unique period of an officer's career.

Commanding an Air Force Squadron was written during the 1989-90 academic year, at the end of the cold war and before Operations Desert Shield and Desert Storm. By the time you may read this book, the Air Force's organizational structure may have changed at various levels, including at the squadron level. Don't let this detract from the book—the command and leadership principles described herein have stood and will stand the test of time and change.

A great many people deserve thanks for their help with this book. The many former and in-place squadron commanders, who gave freely of their time and thoughts during hours of interviews I conducted gathering material for the text. My classmates of the National War College, class of 1990, who gave me constant, enthusiastic encouragement for this project and consistently reminded me of the need for this type of book. Dr Fred Kiley, head of the Research Department at the National Defense University, for his time, patience, excellent guidance on writing style, and tireless efforts in finding a home for the manuscript—all of which tremendously helped this math major actually produce a book. To everyone at the Air University Press for their hard work in turning my manuscript into a finished product—Dr Glenn Morton, my editor, and the entire staff of the Production Division. Finally, my wife, Linda, and our two daughters, Michelle and Nicole, for their love, support, help, and understanding—all of which were absolutely critical to my ability to complete this work. To all those mentioned above, and many more, my heartfelt thanks for your assistance to me with this book.

TIMOTHY T. TIMMONS, Col, USAF
Research Fellow
National Defense University

Introduction

Throughout my military career, I had frequently heard squadron commanders say their job was the best in the Air Force. I first remember those words coming from Col Charles Geer, commander of the 22d Military Airlift Squadron at Travis AFB, California, during remarks at his farewell party in 1977. I respected this officer very much, so I made note of his comments and tucked them away for future reference.

Years later I learned the true meaning of these words for myself, up close and personal. I had the honor of commanding the 450th Flying Training Squadron at Mather AFB, California, from 6 March 1987 until 15 March 1989. Commanding an Air Force squadron is the best job in the service—I fully concur with those who have spoken those words before me.

Why is it the best job? I explain it this way: commanding a squadron is the only time in your career when you are simultaneously close to the mission and in command. Any duty below squadron command may be close to the mission, but the mission is being accomplished somebody else's way. Duty above squadron command may involve command and doing things your way, but you are too far removed from the mission and the people. Squadron command is that one point in time when you have the best of both worlds!

Commanding a squadron is not an easy task—a commander is fully involved in every aspect, nook, and cranny of the unit. The duty encompasses every second of every day the commander holds the unit flag; he must eat, sleep, and drink his squadron, full time. The commander's job is unique in the unit; no other job approaches it in scope—the commander exists on his own "turf."

Why a book on commanding an Air Force squadron? Because I believe (1) the duty is special and hence the topic is worth the effort, (2) given the opportunity to do a long research project, I wanted to produce a product that would be of use to many people, and (3) I feel most new squadron commanders are not well prepared for command when they get it. The Air Force does not have a pre-command course like the Army or the Navy, and I have found little literature concerning squadron command that did not cite a lot of regulations or just quote many senior officers.

One purpose of this book is to help a new squadron commander get acquainted with the roles and responsibilities that go along with the job—written from the viewpoint of those who have either just completed command tours or are still in command. Another purpose is to give a unit commander a single source from which to gather ideas on many topics related to squadron command. A final purpose is to give anyone who desires it the opportunity to read how many junior colonels and senior lieutenant colonels viewed this challenging assignment. To gather material for this writing, I interviewed a large number of officers who had just come off squadron command tours, plus several more who were still in command. Like myself, many of the officers were attending senior service school. Those interviewed represented a wide cross-section of squadron types and major commands. Each interview consisted of 20 questions that took the commander through his command tour in a chronological and topical manner. I am indebted to those who gave their time for the interviews—each was unique and greatly added to the book.

This publication is not a regulation or manual, nor does it quote many regulations and manuals. It does not express the official thoughts or policy of the United States Air Force. It is not meant to be directive in nature—all ideas and thoughts presented represent recommendations and suggestions gathered from a vast group of experienced officers. It is certainly not gospel!

This book does not attempt to cover every possible subject relating to squadron command and certainly does not cover every possible situation when discussing any one particular topic. When reading this volume, some commanders may find that many of the topics directly apply to themselves and their squadrons. Other commanders may find few topics in this book that apply to their units.

I've tried to write this book in very plain English, with few ten-dollar words. It is organized both chronologically and topically in an easy-to-follow, practical manner. The major topic groupings deal with mission, people, and daily leadership/management issues. At various spots throughout the book, I've included a series of one-liners under the heading "best advice." These words of wisdom are direct quotes from the commanders I interviewed in response to the question, "What pieces of advice would you give a new commander, based on what you feel are the most important aspects of commanding an Air Force squadron?"

Hopefully, you'll find the book equally easy to read front to back or to serve as a reference document for ideas on certain topics. *Commanding an Air Force Squadron* is often anecdotal in nature and contains many examples of how commanders handled various situations. My goal in writing this volume is to produce a document that can be of use to a wide array of officers for some years to come. If a commander, new or experienced, can pick up this book and get just one or two ideas that will help him or her improve the squadron, then I've accomplished my goal.

Enjoy.

Chapter 1

Critical Months

Starting off on the right foot as a squadron commander is critical to the overall success of a command tour. This chapter starts at the "beginning" and discusses the timeframe from when an officer is officially notified of being selected to command a unit until he or she has been in command for about three months. Experience has proven that a commander who is "ahead of the ballgame" during this period will stay "ahead" throughout his or her tenure.

Before Taking Over

Certain moments in an officer's career stand out as unique or special. Receiving the news that you've been selected for command is certainly one such special moment. Most officers who have commanded an Air Force squadron will remember clearly how they received the good news. During my own tour at Mather AFB, I must have talked on the telephone thousands of times to hundreds of people about a multitude of subjects. All of these calls have long faded in my memory except one.

About 1500 one day in early February 1987, my secretary at the wing standardization/evaluation division summoned me to the phone, "Boss, the wing commander is on extension 4140 for you." Wondering what the subject might be and remembering that the wing commander had left for command headquarters at Randolph AFB earlier that morning, I answered. He got right to the point, "Tim, this is Colonel Collins, are you ready to take over the 450th?" I replied in nanoseconds—"YES SIR!"—and then realized I had probably shattered his eardrum. He congratulated me and explained that the wing vice-commander would inform the key wing staff of the change of command. By the end of the conversation, my entire division knew what had happened, and I was mobbed by well-wishers. They presented me a 450th Flying Training Squadron scarf, complete with my name and "Squadron Commander" embroidered in script. They had prepared this memento "just in case." I was on top of the world—now I'd be able to find out for myself if all the great things I had heard about squadron command were true.

Most commanders will tell you how busy they became between selection for squadron command and the actual time of their change of command. Time management is critical during this period, because you are pulled in many different directions. The bottom line is that you must establish your priorities between the responsibilities of the current duty and gathering information about the new job. As the change of command approaches, it's wise to slowly phase out of the old job and devote more time to learning the new one.

All commanders and former commanders I talked with agree that it is best to make a clean break with your prior job when taking on the new command—to finish all the taskings and paperwork associated with your old duty before the change of command. Once the change of command occurs, your new duty demands 100 percent of your time. I know this was true in my case. As chief of the wing's standardization/evaluation division, I supervised 17 flight examiner pilots and navigators plus an administrative staff. My departure therefore generated several officer and enlisted performance reports, all of which I wanted complete before the change of command. I was fortunate because I was staying right on the same base, not going to PCS anywhere. So, I had ample time to give these reports the attention they deserved—time that would have been divided with PCS concerns had I been reassigned elsewhere. I had also been working a couple of assignments for my officers and had time to firm those up while they still worked for me. But not everyone is so lucky.

One commander told me that he learned of his selection for command while assigned to the Air Staff in the Pentagon. He had to prepare for a short-notice overseas PCS in addition to concluding some important Air Staff projects. Because of these circumstances, his last two months in the Pentagon were the busiest two months of his career, leaving him with little time to gather information about his upcoming command. He gave his Air Staff work his best effort and assumed the new command with less information than desirable. This dilemma is a common one for officers coming from staff assignments in major command headquarters or the Pentagon. If faced with such a situation, you may seek out someone at your current location who has recently arrived from your future base. Additionally, the telephone may be your best vehicle for information gathering in this scenario. Most officers going overseas to take command of a squadron must complete performance reports—must have them absolutely final—before departure. It is next to impossible to make last-minute changes on reports from 9,000 miles away, and it is a poor practice to leave behind a stack of pre-signed forms. Completing these reports early is number A1 priority before departing.

PROVERB FOR COMMAND

"Get comfortable with the squadron's mission before taking command."

I asked many former squadron commanders what kind of preparation for their new unit they had accomplished before taking over. The consensus was that, within the individual circumstances and time available, a new commander-designate should concentrate on the following four items in order of priority: learning the squadron's mission, learning about the squadron's personnel, meeting the chain of command above him, and understanding the role of other units on the base.

The Mission

Lt Col John Bell, chief of the wing training division and KC-135 navigator at Robins AFB, Georgia, was selected to command the

19th Field Maintenance Squadron at Robins. Although familiar with the aircraft, he had never been a maintenance officer, and like many rated officers, he had little experience working with enlisted personnel. John knew he had to learn the business from scratch, so before he took over, he went to the most knowledgeable maintenance personnel on base—the senior enlisted supervisors in his new squadron—and had them teach him the mission. Not only did he learn a lot about his new duty, but he also began establishing solid relationships with some of the key people in the unit. Besides spending time with the officer he was going to replace, John consulted with other maintenance squadron commanders on base to learn how his squadron fit into the overall maintenance picture.[1]

Sometimes taking command of a new squadron will mean requalifying in a weapons system. Lt Col Bob Lawrence had previously been a C-141 instructor pilot and an action officer on the Air Staff when he was selected to command the 86th Military Airlift Squadron at Travis AFB, California. Bob was able to PCS fairly quickly and begin his requalifying program at Travis. Convinced that a flying squadron commander should be highly competent in his weapons system, he concentrated his efforts in this direction, staying out of the squadron except to talk with the current commander. Bob spent many hours hitting the books and preparing for his flight and simulator missions. When he took command, he was more than fully competent in the primary mission, and there's no substitute for that in the eyes of subordinates.[2]

Often, the requalification program necessitates a TDY school, complicating the new commander's decision on how to allocate his time. Lt Col Mo Blackmore also placed a great deal of importance on being a technically competent commander. Stationed in the Pentagon when he learned that he would take command of the 509th Strategic Missile Squadron at Whiteman AFB, Missouri, he had to requalify in the Minuteman II, which meant a TDY at Vandenberg AFB. Mo decided to PCS to Whiteman first, then travel on to Vandenberg. This decision enabled him to meet the current commander, the operations officer, and the senior officers on the base before his TDY. It also enabled him to settle all the details for his family's move to Missouri. Once all this was accomplished, he was able put 100 percent of his effort into the requalification program, then assume command at Whiteman fully confident in his personal mission competence.[3]

Often an officer is already assigned to the unit he is going to command and hence already mission-qualified. Lt Col Ron Love, as operations officer of the 11th Aeromedical Airlift Squadron at Scott AFB, Illinois, was slated to take over the same unit. Once selected, Ron began to pay particular attention to the commander's duties and made the most of opportunities to serve as acting commander during the incumbent's absence. Being fully qualified and very knowledgeable about the unit's operations enabled him to begin learning the squadron's mission from the commander's perspective.[4]

The People

Gaining knowledge of your new unit's personnel is second in importance only to gaining knowledge of your new unit's mission. This doesn't have to be a long involved process; you don't have to meet and talk to every member of the new squadron. When Lt Col Ray Chapman left the Pentagon to command the 86th Flying Training Squadron at Laughlin AFB, Texas, he had to requalify in the T-38 at Laughlin. While in this necessary training, Ray decided to sit back and just observe how squadron members went about their daily duties. This period of concentrated observation gave him very good insights into the strengths and weaknesses of many of the unit's instructor pilots.[5]

Another former commander told me that before moving PCS, he asked his new squadron to send him a roster of all unit personnel and their families. Before he arrived at his new base, he attempted to learn as many names as he could so that when he first met the squadron members, he knew their first names, their spouses' names, and something about their families. It is not essential to go to such an effort, but—time permitting—it certainly sends a good message to the new squadron.

Most commanders get a chance to sit down with the officers they are replacing and discuss mission and people. Such conferences are highly recommended if at all possible. But one caution: while it's good to hear the outgoing commander's detailed rundown on all unit personnel, you should not prejudge everybody based on the outgoing commander's comments. Make a mental note of the comments, but wait and see how the personnel perform under your command.

Some situations allow the new commander to learn about the squadron's mission and people at the same time. Lt Col Tom O'Riordan was the first commander of the 37th Bomb Squadron (B-1B) at Ellsworth AFB, South Dakota. He attended B-1B Combat Crew Training School at Dyess AFB, Texas, with a large number of the officers who would be assigned to this new unit. He used this opportunity to develop solid relationships with his crewmembers and to discuss squadron mission philosophy and goals. Upon arrival at Ellsworth, Tom already had a big start towards forming his new squadron.[6]

The Chain of Command

Command relationships are important in every Air Force unit. Many commanders make a special effort to meet with their new bosses before the change of command. That way, the new commander knows from the start what is, and is not, important to those up the chain of command.

Remember how Lt Col Mo Blackmore PCS'd early to his new base before going TDY for training? Part of his plan was to meet his chain of command early to establish a solid groundwork for the future. Another lieutenant colonel told me that he had made a mistake by not taking the initiative in scheduling such a meeting. He felt a little behind the power curve because he didn't get a chance to sit down with his new boss until three weeks after his change of command. A former SAC unit commander told me that because the officer he replaced had been fired, this meeting with his new boss was important since it contained guidance concerning squadron problems. Scheduling this meeting early enabled this commander to begin his tenure fully aware of what had gone wrong before and in what direction he needed to point his new squadron.

Even if you are stationed at the same base as the new squadron, you will want to talk to your new chain of command about their views on your new duties. This holds true even if you are not changing supervisors—your boss's view of a squadron commander will be different from his or her view of whatever duty you presently hold.

The Base Environment

My own preparation for taking command was fairly easy. I had been on base for two and one-half years, had held two wing staff supervisory positions, and had been the operations officer of the squadron I was going to command. My predecessor and I were good friends and usually ran together three or four times a week. I was comfortable with the unit's mission, its people, and the chain of command on the base. However, I was uninformed concerning many of the other units and agencies that I would now deal with as a commander. So I took the initiative and went visiting. Places like our CBPO gave me a complete tour and package that told me everything I'd ever want to know about their work. Their people were eager to explain their job to me and tell how they could assist me in my duties. I then turned things around and asked them how I, as a squadron commander, could help them. I paid close attention to their answers and later did everything I could to follow through. Getting to know the lawyers at the base legal office early on also proved beneficial during my command, as did talking to the key people at supply, hospital, civil engineering, and security police.

Most commanders will verify the need to be sensitive to the situation of the departing commander. It's probably a good idea before the change of command to visit the new squadron only when invited by the outgoing commander. The new commander's presence can inhibit the departing commander's ability to lead, by casting him as kind of a lame duck. One officer said that every time he went into his new unit, a small crowd would gather, people looking to get some early "face time" with the new boss.

One situation I was sensitive to was the squadron's going away party for the outgoing commander. Even though he and I and our wives were good friends, and I had served as his operations officer for a while, I did not attend the function. I felt that this was his party and my presence there would only detract from his night. I even went so far as to contact the officer in charge of the function well ahead of time to explain my feelings. In retrospect, I am sure this was the proper decision.

The next assignment of the outgoing commander may make a difference in how you prepare and ease into your new duty. One former F-15 squadron commander told me that he was very sensitive to the fact that the officer he replaced became the assistant deputy commander for operations and remained on station for six months before going PCS to another location. This situation made things somewhat difficult for the new commander, because the two officers had different operating styles.

One officer who was going "across the street" to take command of a similar squadron said that he was very careful what he said when asked what changes he had planned for his new unit. He felt the need to be totally positive towards his new unit and to avoid any comparisons with his old squadron. He also said that he felt the worst thing he could do before the change of command would be to talk about "turning his new unit upside down." Therefore, even though he had some firm ideas on future changes, he kept them to himself until he was in charge—an attitude that can certainly do no harm and that leaves the new commander in a position to act more confidently once he assumes command.

What about the "art of command"; how do you "study up" for your command duties? In the Air Force, a squadron commander usually learns how to command by commanding. We don't place nearly the emphasis on command training as does the Army. LTC Steve Rippe commanded an infantry battalion in Aschaffenburg, West Germany. He was informed about his command a full 18 months before he was to take over. During this time he attended three separate Army courses to prepare him for his command duty. These courses stressed the nature of command and included a full-blown command field exercise.[7]

Chances are that a new Air Force squadron commander will not receive any training close to what Steve Rippe received. However, if you find yourself preparing for your command duty and feel uneasy about the nature of command, there are some good books out that can give you some insight on this topic. A couple I'd recommend are *Taking Charge: A Practical Guide for Leaders*, by Maj Gen Perry Smith, USAF, Retired, and *Small Unit Leadership*, by Col Dandridge Malone, USA, Retired. A new book, *Company Command: The Bottom Line*, by Col John G. Meyer, Jr., USA, though written for Army captains, is a treasure chest of tips, war stories, and anecdotal advice that can help any squadron commander, and, in these days of jointness, is a useful book for your captains to read.

While we are talking about preparing for command, let's not forget your family. Their lives can change substantially as you take over your new responsibilities. Your wife will inherit some work, also, and she won't get paid one thin dime! One former SAC squadron commander told me that he asked a lot of questions upon arriving at his new base concerning any spouse responsibilities so that his wife would not get blindsided when she arrived with the kids just before the big day. It's a real good idea to explain fully to your wife what you've learned about your new squadron as you prepare for command. Don't forget your children either. If they are old enough to understand, talk to them about your new duties and how your new position may affect them.

Normally the squadron and the outgoing commander take charge of the actual change-of-command ceremony. Customs for this day vary greatly by command and base. I recommend that an incoming commander ensure that he is kept abreast of details as they are planned. A C-130 squadron commander related that his incoming change of command took place six hours after his arrival at his new overseas location, in flightsuits in a hangar. Being aware of this well in advance, he was fully prepared and avoided running around at the last minute trying to set up a new flightsuit.

New commanders usually have an opportunity to address the audience at the change-of-command ceremony. Here are a few tips:

- Be brief.
- Thank the many people who have helped in the past.
- Lay out a general idea on where you want to lead the unit.
- Add a quick anecdote about your relationship with the outgoing commander.
- Tell the new unit how proud you are to join their squadron.
- Be upbeat and positive.

After the change-of-command ceremony, it's time to enjoy the reception, party, or whatever—the hard work starts the next morning!

Getting Started and Setting the Course—The First Three Months

The first three months of a new squadron commander's tour set the tone for his entire tenure. Most commanders agree that this time is somewhat like a presidential "honeymoon" period, in that everyone generally waits a few months to see how you handle your new duties before passing any judgement on your leadership ability. To start off on the right foot as a new commander, you may want to consider accomplishing the following during this 90-day period:

1. Understanding your new position and how it affects others.
2. Cementing your mission knowledge and expertise.
3. Getting to know your personnel.
4. Ascertaining the "health" of your unit.
5. Learning how your squadron fits into the wing's mission.
6. Determining the direction you want your unit to take under your command—the squadron's focus.

Your Position

You will begin to understand something about how your new position affects others the first morning you walk into the squadron after your change of command. It may be obvious—some units traditionally call the building to attention when the commander arrives in the morning. It may be very subtle—unit personnel will talk to you and even look at you in a new manner. Regardless of how you find out, one thing you will quickly understand about your new position is that you are no longer "one of the guys." This realization can be a big shock. It may take you about three to four weeks to accept the fact that the very nature of command has separated you from everyone, and this separation is for the good of all concerned.

Another fact a commander has to accept quickly—command is a 24-hour-a-day, seven-day-a-week job that allows for less personal freedom than most other duties. I found this true—in many ways I was married to the squadron. First, the squadron needed to know where I was located during the duty day. A piece of advice—be totally honest here with your people and instruct your squadron to be totally honest about your location. If you are going off base to run some personal errands, say so, you're the boss.

> **PROVERBS FOR COMMAND**
>
> "Its probably the first job you'll have where you must distance yourself from your friends."
>
> "Commanders underestimate the threat they are."

Second, my freedom to gather information concerning daily occurrences in the unit was significantly different from that during my previous duties. As a new commander, I found that people were less willing to come to me with bad news or problems until I could convince them that when I said "open door," I really meant it. Four TIPs here to make this work: never shoot the messenger, never pass the buck when a problem comes to you, even when you've got a million things going, take time to talk to your people, and concentrate on what they're saying as if they are the most important people in the world.

Third, the demands placed on my time by meetings, counselings, and paperwork restricted my daily schedule. I would have preferred to fly as often as when I was an operations officer, but this simply wasn't possible. This is a tough pill to swallow. Command may infringe on some of the more enjoyable activities you like to pursue—it comes with the turf! Finally, how I conducted myself on a day-to-day basis affected my unit. I stressed being upbeat and positive, and it rubbed off on the unit. When I received bad news, got chewed out from above, or just plain had a bad day, I was careful in how I reacted publicly to the situation. My having an off day should not force everyone else to have the same type of day.

One MAC squadron commander told me he stressed professionalism in his position—all the way down to how he kept his office. When he took over, he felt the upkeep on the squadron building was something "less than perfect," and he noticed that his predecessor's office always looked like "tornado alley." Without saying anything, he worked at organizing and keeping his office in a neat and presentable condition. Any papers on his desk were always stacked in an orderly fashion, his conference table was free of papers, dead plants were removed, and chairs were arranged in a specific pattern—he went "overboard" for a reason. Slowly, but surely, this rubbed off on the rest of the unit, and the squadron building began to look professional also.

Besides understanding how your new command position affects unit personnel, you will likely also notice changes in your peer group and your relationship with your boss. One former commander found his new peer group to be the other squadron commanders on the base. Because of the uniqueness of squadron command, the only other officers who could fully understand and appreciate his responsibilities and span of authority were other commanders. That didn't mean he was no longer close to his old friends, but when seeking advice or information on certain topics or issues, he now started with other commanders. Similarly, although his immediate supervisor did not change when he took over a squadron, their relationship did change. Before, as a division chief on the supervisor's staff, he had enjoyed a good working relationship with the boss and was an integral part of the "deputate team." However, as a squadron commander, he felt an even closer relationship with his boss because he now had more latitude to run his unit as he saw fit, and at the same time was more responsible and accountable for a far greater range of tasks. This is one difference between being "a boss" and being "a commander." I think Lt Col Steve Lorenz put it best: "As a squadron commander you are the focal point of a living organism"—everything revolves around and points towards your actions. You are the vital link between all other wing organizations and every single individual in your unit.[8]

This being the case, once you understand your role and the idea that you now work in a "glass house," how do you act in this environment? The best solution here is perhaps the simplest one—you don't change! One of the best pieces of advice I ever received came from my wing commander, who told me not to change how I worked because I was now a squadron commander. He said that what I had done to get me to this position would keep me in good stead during my command. He also told me not to attempt to command like my predecessor, but to run the squadron as I saw fit. When I thought about it, this advice made perfectly good sense—the powers that be obviously liked my performance or they wouldn't have given me a squadron, so there was no reason to change and all the more reason to continue to be myself.

> **PROVERB FOR COMMAND**
>
> "Don't change—you wouldn't be in command if you didn't deserve it."

The Mission

Gen Curtis E. LeMay once said, "Three things are important, the mission . . . the mission . . . the mission!" I think all commanders will agree the bottom line for any military unit is its ability to perform its mission in peacetime or wartime. Your squadron's ability to accomplish any tasking can greatly depend on your personal expertise or knowledge of the mission. Although you may have an operations officer or chief supervisor whose primary duty is specifically to oversee the unit mission, as commander you are responsible for the mission. Every single commander I interviewed agreed that the best method to command a unit's mission is simply to "lead the way and be out front" as much as possible. All of them ranked understanding the mission "cold" at the top of any list of a commander's priorities. As Colonel O'Riordan put it, "Solid technical competence on the commander's part gives him the freedom to lead."[9]

Sometimes this means completing an upgrade or special qualification program. If so, completing the program early in your tour as commander will prove beneficial. Leading the way can also mean occasionally taking on some of the less pleasant taskings such as alert duties, Friday night flights, or even tedious meetings where you could have sent someone else. Colonel Blackmore periodically pulled a 24-hour Minuteman II tour in the hole. He didn't have to do this, but to him leading the way was personally doing the same mission he asked his people to perform.[10]

A commander also needs to understand that some circumstances will prevent him or her from performing every mission that his or her people perform. A rated officer who has taken over a maintenance squadron cannot reasonably be expected to possess the same knowledge and expertise as the senior supervisors. Realizing he was in this exact circumstance, Colonel Bell gained his knowledge of the unit mission by constantly getting out of his office and "talking mission" with everyone in his squadron—from his supervisors and superintendents on down to his young airmen on the flight line. He felt it was important to get out early during his command and learn the mission from his troops, at all times of the day and the night.[11]

Colonel Lawrence had a solid working knowledge of the C-141 airlift business when he took over his unit at Travis. What he needed to understand better was specifically how his particular squadron performed its mission. To this end, he requested a series of briefings from his unit's sections where he let his people teach him the local specifics.[12]

A solid working knowledge of the mission means knowledge not only at your level as commander or your subordinates' level but also mission knowledge as it relates to your boss. The best way to gain this type of knowledge—how your boss views your mission—is talking to him or her in some detail about the subject. One commander related how she scheduled herself on her boss's calendar the first week after taking command just for the purpose of learning exactly how he viewed the mission of her squadron. She ranked this as one of the smarter moves she made as commander.

PROVERBS FOR COMMAND

"Technical competence gives the commander the freedom to lead."

"Know your technical skills to start with, people look up to you."

Your People

Most commanders will tell you that knowing your people ranks on the same level as knowing the unit's mission. Not only does the commander have to be knowledgeable on the strengths and weaknesses of unit personnel relative to mission accomplishment but also relative to their Air Force careers and, to some degree, their personal lives. Knowledge of each individual's mission ability will help the commander fully understand the squadron's overall capability. Knowledge of individual strengths and weaknesses relative to a successful career and individual desires is important because the commander is the unit's career focal point. Finally, a general knowledge of his or her people's personal lives will allow the commander to avert potential problems that can negatively influence mission accomplishment. None of this is an easy chore.

Firsthand observation is the best way to get to know your people's ability to perform the mission. Many flying squadron commanders make it a point to fly with as many of their crewmembers as possible. To augment this type of observation, commanders actively seek the opinion of key squadron supervisors regarding individual performance; the commander cannot be everywhere. One commander told me he could get a feel for people's ability to perform the mission just by talking to them about it. He never specifically quizzed individuals about their duties, but rather preferred to walk around his squadron and get into group discussions about the mission. This not only got him out of the office and around the unit, but it also enabled him to see how unit personnel interacted with each other.

Commanders place a premium on taking care of their people's military careers. To best accomplish this, a commander needs to know two things—the individual's record and the individual's desires. These two items, coupled with performance, should enable a commander to guide and counsel his people effectively and honestly. Getting a good handle on an individual's record and current performance can usually be accomplished during the first three months of command. Getting to know everyone's desires will normally take longer.

PROVERB FOR COMMAND

"Quickly find out what makes the squadron tick."

One former SAC squadron commander explained how he accomplished this task. When he set out to learn the background of his squadron members, the first thing he decided upon was exactly what things he needed to know from their records. Next, he went to the CBPO to look at the records, usually for about an hour at a time, until he eventually had reviewed each squadron member's (officer and enlisted) folder. While there, he made out for each person a chart that included the information he wanted readily available—various service dates, past assignments, past performance reports and notes on other items such as photos, Form 90s, and educational and PME background. He then ordered up a personnel brief on each squadron member from the CBPO and made up a notebook that contained the brief and his chart on every member of the squadron. He kept this notebook in his desk drawer, next to the telephone, at all times.

Finally, to show his people he was interested in their careers and performance, he and his operations officer revamped their squadron manning boards. Both boards were your basic wiring-diagram depiction of the unit organization with names of every squadron member filled in. On the commander's board were added color-coded symbols indicating each person's date of rank, educational level, and PME (squadron, intermediate, senior) level. The operations officer's board had symbols depicting individual mission qualifications (instructor, step grade) and upgrades. The offices were arranged such that with doors open, the commander could talk on the telephone or to someone in his office and read both boards.

Once the squadron personnel saw that the commander was actively interested in their careers and performance, they took an added interest themselves. The boards were not there to promote competition among squadron members, but rather to give the commander a better view of each individual and the squadron as a whole.

Unit "Health"

Getting to know individuals in your new squadron is usually easier than getting a firm handle on "unit morale." Many commanders tell me they believe the best way to ascertain unit morale is to talk to as many people as possible in their workplace, seeking out those certain "key" individuals who always have a grasp on what's going on. One commander related how soon after taking command of his unit, he had a "gut feel" that there were problems with his unit's morale. He went to a fellow, highly experienced squadron commander for advice. This other commander was well aware of the problems and how they had developed. He offered good advice on how to remedy the situation.

Many times a new commander will inherit a squadron that performs its mission well and has high morale. Shortly after taking over such a unit, one lieutenant colonel told his entire squadron he knew they were good and that he would adapt to the unit first before making any adjustments to fit his style as time went on. This was very well received and he quickly gained the confidence of his supervisors. Most former commanders agree that changing things right away just for the sake of change is not a good idea.

Other new commanders inherit squadrons in trouble. I learned how one commander took over a squadron with $8.00 in the unit fund, without a commander's call in eight months, and facing a major mission change in the next six months. Believing he had no

Proverb for Command

"Don't go in with bugles blaring, but leave with them."

"90-day honeymoon period," he acted quickly. After obtaining some basic guidance from his boss, he held a commander's call where he laid out some specific goals for the squadron and committed himself to get involved personally in each member's career, something his predecessor had not done. He then stated that he wanted to meet with all the senior supervisors in the squadron to plan out courses of action to attain the stated goals. The severity of the problems forced him to act quickly, but his method of approaching the situation won him the confidence and respect of his people.

Some squadrons may not be in severe trouble but may be "down" for one reason or another. I heard of one squadron that was in great shape except for its facility. This squadron was functioning well and unit morale was high, but many members felt the building was a disgrace. So, the new commander put this on top of his priority list, worked closely with the civil engineering commander and finally initiated a good self-help program that turned his building into one of the best on base. This improved unit morale and positively influenced mission performance.

The Wing's Mission

No squadron in the Air Force can successfully perform its mission alone; every unit requires help or support from other units and agencies on the base. Gaining an early understanding of how the squadron fits into the overall wing mission will help the commander lead his or her unit. Many commanders tell me that their indoctrination on this subject began during their first meeting with their wing commander and continued as they met the other unit commanders on the base.

One commander related how he visited most every unit on base during the first three months of his command. Some visits were just courtesy calls and none lasted over an hour. These were designed to give him a working knowledge of the other units and how his squadron interacted with them. He found this knowledge very helpful during the next two years, especially when he encountered specific problems and knew exactly where to turn for help. He also related that the knowledge of how the wing functioned helped him put his unit's mission in proper perspective and helped him better explain certain taskings to his troops in terms of the "big picture."

Setting the Direction

Towards the end of these first three months of command, most new commanders have gained a good understanding of their squadron's mission, people, and place in the wing. At this point they should begin thinking about setting the direction they want their squadron to follow. A couple of former commanders told me they felt it necessary to develop and implement a game plan quickly because any long delay would result in their never personally realizing any results.

One former commander of a very selectively manned unit emphasized what was important to him and set the squadron in a direction to do things his way. The squadron had been doing a good job in the past, but he felt that they were not being responsive to the needs of the major command. He also sensed that the squadron members felt they were in more of a "fraternity" than a military unit. Under these circumstances he set down fairly concrete directions to make the squadron more accountable to the MAJCOM and to change the attitude of the personnel. At first, these changes were not popular within the squadron and there was some initial resistance, but the new commander stuck to his guns and ended up with the support of everyone.

A former SAC missile squadron commander related how his main thrust was to establish pride within his unit. The unit had experienced some problems in the past but had been doing much better in the months just before he took over as commander. However, they were used to being "in the basement" and seemed content with their status on base—inertia had set in. To promote pride and self-respect, the commander specifically showed the unit members how they now compared favorably with the other units on evaluations and other areas of mission accomplishment. He finally convinced his people that they deserved better, and that the best way to rise to the top was to establish unit pride and work hard at being recognized for their accomplishments. This direction worked—the squadron was recognized as the best in the wing the next year.

Another commander of a training squadron related that he would first get his squadron members to "buy into" his overall game plan before formally instituting changes. Early in his command he recognized a span of control problem and felt it wise to increase the number of squadron flights. Instead of dictating this change, he asked the squadron members for some advice concerning the possible organizational change. He already knew the answer would come back positive because it meant more flight commanders and other supervisory positions, but he wanted the squadron to feel a part of the decision. He also wanted to know if there were any potential problems with these moves that he might have overlooked. The unit personnel were very much in favor of the moves, and after receiving the positive feedback, the commander instituted the changes that improved mission effectiveness and unit morale.

Many commanders set the direction for their squadron by what I call "focusing" the unit on certain priorities. I used the word *focus* a great number of times when talking to my squadron, mostly in relation to ensuring that they concentrated their efforts on the unit mission. As a navigator-training squadron commander, I wanted our predominant focus to be on student training, the basic mission. I strongly felt this primary focus should be far above all other efforts of the squadron, and I wanted training to occupy the vast majority of our resources. This wasn't easy in a training environment where many other wing and command programs occupy much of the commander's, operations officer's, and flight commander's time, but I insisted that the actual student training remain on top of our pile of taskings. I preached this philosophy to anyone who would listen and also placed the mission first in my daily activities. This meant we might fall a little behind elsewhere, but I had set a specific path for my unit to follow.

Colonel O'Riordan was able to establish a new B-1B squadron and had a free shot at setting unit focus and direction. His entire unit worked on a "Tiger Plan" that would serve as the basic written plan for the broad spectrum of squadron activities. Within this plan, he emphasized three items—mission, safety, and people recognition—this set the direction and priorities for the new squadron.[13]

Some new commanders feel the need to continue the squadron in the current direction with only a few minor adjustments. One MAC unit commander wanted to emphasize the mission a little more and make the work environment more comfortable for his C-141 crews. Basically, he strove to make flying enjoyable again and then used the mission to build unit cohesion. To accomplish this, he stressed the positive, fun aspects of military flying and de-emphasized, where possible, the more unpleasant aspects that go along with the mission.

Other new commanders tell me they picked specific portions of the mission to emphasize early during their command tour. Colonel Chapman focused his T-38 squadron squarely on safety. Flying sometimes more than 200 sorties a day, many of them student solos, Ray felt the need to review mission safety continuously. As the commander, he emphasized "stepping back, getting out of the forest, climbing up on the hill, looking around, and evaluating what's going on concerning flying safety." There was no doubt where Ray Chapman was focused.[14]

Col Bill Welser, who commanded a C-141 squadron at McGuire AFB, New Jersey, focused a portion of his attention towards the conduct of his unit's internal business. One of the first changes he made was to give his supervisors a chance to supervise and not be micromanaged by the commander. He emphasized to his entire unit the concept of solving problems inside the squadron as opposed to taking them up the chain to his boss. When faced with difficulties, he urged everyone to "be part of the solution, not part of the continuing problem." These directions were all designed to improve the unit's internal operations.[15]

Sometimes the commander sees the need to direct his squadron outward. A former C-130 squadron commander at Little Rock AFB, Arkansas, needed to enhance his young unit's self-confidence. To achieve this, he developed a large public relations effort to get the accomplishments of his unit and its members recognized by the wing and local community. He used the base and community newspapers not only to publicize the achievements of unit members but also to educate the public on his squadron's operation. Once this program was fully implemented, he saw a marked improvement in the squadron's confidence and performance.

Most of the examples I've just discussed were fairly broad in scope; some commanders may want to be very specific when setting direction and goals for their unit. Colonel Rippe set three goals for his infantry battalion, each critical to mission accomplishment: (1) be able to roll into their tactical assembly area flawlessly, (2) meet all high Army gunnery standards, and (3) meet all critical standards on every inspection. He set these goals shortly after taking command and enacted a few internal changes to help achieve the goals, all within his first five weeks. Steve is very action-oriented and believes that the time to fix a problem is now.[16]

Two of the officers I interviewed had already helped set their squadron's direction while serving as operations officer and maintenance superintendent. They did not feel a need to set any new direction since they were perfectly happy with the current focus of the squadron.

Another commander had a much tougher road to travel. He was very happy with his unit's current direction, but the Air Force was about to significantly alter the unit's mission. His focus then turned to planning for the upcoming mission change, ensuring it came off smoothly with no loss of readiness or effectiveness.

The toughest road to travel is setting the direction of a unit that is really in trouble. Often a new commander discovers many reasons why the squadron is failing. A few guidelines for this type of situation: (1) seek out and emphasize the good points to give people a hope for the future, (2) believe your instincts and don't be afraid to make significant changes, (3) listen to your boss's guidance, and (4) do not hesitate to fire people when necessary, but do so as a last resort.

> **PROVERBS FOR COMMAND**
>
> "Decide your focus and how to get there."
>
> "Provide guidance on fundamental goals and prioritize these goals."
>
> "You need a general philosophy and overall goal."

Within the first three months of their command tour, all of the commanders mentioned in this chapter learned that (1) they were the focal point for the unit, (2) all squadron personnel looked to them for direction, and (3) a squadron needs a leader out front, not a manager behind a desk.

Notes

1. Lt Col John Bell, interview with author, Ft NcNair, D.C., 19 September 1989.

2. Lt Col Bob Lawrence, interview with author, Ft McNair, D.C., 9 November 1989.

3. Lt Col Mo Blackmore, interview with author, Ft McNair, D.C., 9 November 1989.

4. Lt Col Ron Love, interview with author, Ft McNair, D.C., 1 November 1989.

5. Lt Col Ray Chapman, interview with author, Ft McNair, D.C., 19 September 1989.

6. Lt Col Tom O'Riordan, interview with author, Ft McNair, D.C., 10 October, 1989.

7. LTC Steve Rippe, interview with author, Ft McNair, D.C., 2 November 1989.

8. Lt Col Steve Lorenz, interview with author, Ft McNair, D.C., 16 October 1989.

9. O'Riordan interview.

10. Blackmore interview.

11. Bell interview.

12. Lawrence interview.

13. O'Riordan interview.

14. Chapman interview.

15. Col Bill Welser, interview with author, Pentagon, 27 October 1989.

16. Rippe interview.

Chapter 2

The Mission—Top Priority

The squadron's primary mission needs to be a commander's number one priority. There are many factors that affect a unit's ability to perform its mission effectively. This chapter focuses on three mission-oriented topics: a commander's relationship with his peers and his boss, how a commander can build unit cohesion, and an "old favorite," inspections. These subjects may be diverse, but the common thread of the unit's mission runs through all three.

Command Relationships—Your Peers and Your Boss

With regard to unit effectiveness, these relationships are as important as those internal to the squadron. A single squadron is one of many in the wing or base structure—few units can perform their mission in isolation; cooperation among units is essential to success in the military. Furthermore, one of a commander's key jobs is to oversee interactions with activities outside the unit, thus allowing the squadron members freedom to accomplish the mission internally. A sharp commander acts as a combination giant shield and thick sponge—reflecting, absorbing, and filtering external inputs to the squadron, enabling everyone else to go about their business. We'll start with the other squadron commanders on base.

Your Peers

The moment you take command of a squadron your peer group changes and shrinks considerably. Besides no longer being "one of the guys" in the unit, you are no longer one of the guys around base. The full impact of this second change is not as readily apparent as the first change, but sooner or later, you'll get the picture. This change isn't necessarily good or bad, its just part of your new turf.

Lt Col Bob Lawrence explained it best: "The other squadron commanders on base are the only guys who know what you're going through at the moment."[1] Squadron command is unique; you almost have to experience it to fully understand it. Further, unless you're presently in command, you can't appreciate the current issues and problems. Therefore, your new peer group (the other unit commanders in the wing) will be the people you'll most likely turn to for coordination and advice.

Can this group be influential? You better believe it! The extent of this group's influence will depend on many factors, most important of which is how well they all get along with each other. Most commanders related to me how they enjoyed a fairly good relationship with their peers. A few situations concerning personality differences were mentioned, and two of the commanders talked about "that one guy who always has a different agenda!"

As a new commander, you will rarely enter a situation where every unit commander totally agrees on every issue. Rather, there will likely be a lot of give and take on subjects that affect all squadrons. Also, natural rivalries are present in most wings. These rivalries may be a result of multiple units performing the same or very similar missions, squadrons with slightly different missions but using the same equipment or competing for the same airspace, or units who must "borrow" or use personnel from other units. These rivalries or competitions can be healthy for the wing if the squadron commanders can maintain their focus on better mission accomplishment. These rivalries can get ugly if the involved commanders start to take things too personally and focus their squadron on doing better while simultaneously making the other unit look bad—this accomplishes nothing!

The basic organization of some wings creates natural barriers to totally smooth squadron relationships. Two different commanders explained how one of their wing's squadrons trained personnel for the three operational squadrons. The wing dynamics here led to a sometimes adversarial relationship between the operational units and the training unit. The commanders involved eventually solved their squadrons' long-existing differences, but not without a good deal of compromise on each side. The main lesson—the commanders were smart enough to place the wing's mission above squadron-only interests.

Lt Col John Gibeau experienced a similar situation in Florennes, Belgium. He commanded an operational GLCM squadron that worked closely with the maintenance and security police squadrons. When the wing deployed in the field, his operational squadron was in charge of personnel from all three squadrons. Before John took over, the relationship between the three units and their commanders was, at best, OK. During the next year the other two commanders changed, and their replacements sat down along with Colonel Gibeau and worked out many long-standing differences. Finally, through leadership by example, the squadrons' members saw the light and the units began to work well together at all levels.[2]

> **Proverb for Command**
>
> "Never promote your unit to the detriment of another squadron."

The key elements in the last example were the squadron commanders. They decided to work together first and then their units naturally followed suit. What are some of the things the peer group of squadron commanders can do to build a better team relationship? Here are a few examples:

1. Periodic meetings of squadron commanders to discuss current issues and topics are great facilitators of cross-feed communication. The commanders at McGuire AFB get together for breakfast, those at Little Rock AFB for lunch, once a month.
2. More informal get-togethers work well also. Some of the

Travis AFB commanders would meet and relax together on Friday afternoons. Many of the Mather AFB commanders would run together at noon a couple of times a week. The joke around base was, "I wonder what they'll come back with this time after five miles?"

3. One commander explained to me how he and his fellow commanders met once a quarter to discuss upcoming wing personnel moves. If they all concurred on some proposed replacements, they would actively approach their boss with their unified suggestions. Once they even suggested a replacement for a wing job because they all felt the incumbent wasn't performing well. Soon after, the move was made—the boss must make this type of decision, but a group of squadron commanders can be an influential lobby.

4. Party together! Two or three times a year, the squadron commanders and their wives might go out as a group, with nobody else along. The commander who suggested this idea to me said it was pleasant going to a party where everyone there was on the same level, no one below or above you!

Your Boss

As with their peer group, most commanders enjoy a good relationship with their boss. In many cases, a squadron commander's boss has been in charge of a unit and will understand many of the inherent challenges. Also, bosses naturally want to see the commanders succeed; their own success depends on their units' successful performance.

Most bosses or supervisors of squadron commanders will let the commander run the unit as he sees fit; this type of situation is great. However, I've heard of a few bosses who tended to micromanage squadrons—if you get one like this, you may have to walk a tightrope to keep firm control. The micromanagers tend to micromanage only small segments of a unit's operation. The best advice I can offer here is to quickly ascertain what aspects of your unit are likely to get a lot of attention, become a subject matter expert in these hot areas, and do your best to shield your people in any hot area from undue scrutiny.

Another topic concerning your relationship with your boss is feedback. Almost all the commanders I talked with related they received very little direct feedback from their boss about their performance as a squadron commander. This circumstance held true regardless of whether they were close personal friends with their boss or not. Most wish they had received at least a little feedback; a couple commanders said that no news is good news. I can't place a finger on the exact reason for this lack of feedback, but as this is being written, that's the current prevalent mode of operation.

PROVERBS FOR COMMAND

"Your job is to keep your boss out of trouble."

"Get to know and understand your boss's agenda."

"Be aware of the relationship between your boss and the wing commander."

"Work your boss's problems and you won't have any problems."

"It's not up to your boss to get along with you, it's up to you to get along with your boss."

"Know your limits with regard to how far you can advance and push your ideas or proposals; know when to back off."

"Never misrepresent things to your boss and never accept someone's misrepresenting things to you."

"Don't hide anything from your boss."

Finally, a new squadron commander will quickly realize he or she serves more than one boss. The immediate supervisor is likely to be some type of deputy commander, such as the deputy commander for maintenance or the base commander. However, the wing commander is also a boss—the "big boss." Different wing commanders use different styles to run the show. Some will deal or interface solely with their deputies, others will deal mostly with the deputies and some directly with the squadrons, and still others will frequently work directly with the units. A squadron commander's job here is to find out how the game is being played in the wing and adjust to the style. Be smart; if you back-brief your immediate supervisor when necessary, you should stay out of trouble. Also, get a handle on the relationship between your immediate boss and the wing commander. If they are both singing from the same sheet of music, things are relatively easy; if they're not, you may experience some problems you won't be able to avoid.

No squadron is an island that can perform its mission in isolation. Cooperation among units and agencies in a wing starts at the top, with the commanders. If squadron commanders build strong command relationships with their peers and their boss(es), wing mission effectiveness will be enhanced.

Building Unit Cohesion and Morale

"You go to war for freedom and democracy, you risk your life for your unit and your buddies—unit cohesiveness is vital!" So said Lt Col Steve Roser, who was commander, 14th Military Airlift Squadron, from 1987 to 1989. Colonel Roser was not the first to echo these thoughts and he will not be the last. However, as an Air Force squadron commander, he gained a deeper appreciation of these words and their meaning. He also saw how they apply in peacetime—the effectiveness of a unit in performing its primary mission is directly related to its cohesiveness and morale.[3]

Another commander explained it this way: the ability of a unit to perform its mission is not always equal to the sum of its parts; if unit cohesiveness and morale are above average, then the whole (mission effectiveness) will be greater than the sum of its parts; if cohesion is below average, the whole will be less than the sum of its parts. A squadron commander inherently has control over all unit parts—it's his responsibility to add the glue to ensure a cohesive squadron. This is not an easy task—many natural and built-in obstacles stand in the way. For example:

1. A large strategic airlift squadron typically has 20-40 percent of its crewmembers off station on missions at any one time; the only day when all unit members are possibly in town is Christmas!
2. A field maintenance squadron will have a wide diversity of personnel assigned to the enlisted side of the house alone—from chief master sergeants who entered active duty in the mid-1960s to young airmen who came on board in the late-1980s or later! The personal likes, dislikes, and interests of just these two groups are worlds apart.
3. Stateside fighter squadrons often deploy to overseas and/or stateside locations for weeks at a time.
4. A strategic missile squadron always has a portion of its unit on alert, 365 days a year.

Proverb for Command

"You can't mandate pride—you have to build it."

5. Right outside the main gate of most overseas bases "America" just doesn't exist—the culture, language, habits, money, and laws are drastically different.

All the above factors plus many more challenge a squadron commander's ability to lead and build an effective unit in a peacetime situation. Ultimately, if a squadron is not cohesive in peacetime, its chances of pulling together in wartime for the long haul are slim.

Exactly how a commander attacks this challenge will vary with every single squadron in the Air Force—there is no one set solution. In general, the challenge is greater in a large, overseas squadron than in a small, stateside unit, but even these two variables can fool you.

Here now are a couple sets of examples of how commanders approached this task, one set duty related and the other set nonduty related.

Duty-related Examples

Many commanders stress that how a commander treats his people is a big factor in determining unit cohesiveness and morale. Colonel Gibeau relates that upon assumption of command he sensed an underlying split between his officers and enlisted personnel. His own investigation revealed a lack of consistency in the past treatment of officer and enlisted personnel relative to disciplinary matters. The enlisted troops felt they were getting hammered for the same things that officers were "getting away with." Colonel Gibeau put a quick end to this—he discussed the situation at a commander's call and then followed through in the succeeding months with consistent treatment of all unit members. Squadron morale and cohesion immediately picked up.[4]

Many commanders use unit deployments, TDYs, and alerts to build a sense of unity among squadron members. Lt Col Buzz Moseley's F-15 unit deployed five times a year. All officers and enlisted deployed together, and as commander, Buzz felt these deployments gave the squadron members a chance to focus together on the mission without the added distractions of being at home station.[5]

Often the simple act of a commander's personally pulling an unpleasant duty can help build morale. Lt Col Mo Blackmore surprised his squadron by insisting he qualify for and pull 24-hour Minuteman II alerts—down in the hole just like the rest of his crewmembers.[6] Other commanders agree that this type of leading the way can help bring a squadron together.

As a commissioned officer, the commander will naturally be more conversant on officer issues than enlisted and civilian personnel matters. This usually necessitates the commander's working a little harder to understand nonofficer issues. For this reason, a sharp commander will often take a proactive tack towards understanding such topics as OJT, WAPS, the new enlisted evaluation system, and civilian performance reports. A commander who shows an interest in the well-being of his or her enlisted and civilian force will naturally help the unit's morale.

Another way a commander can show interest in his or her people is to visit them in their workplace, individually and in groups. If the members of a section or flight sense the commander's taking an interest in what they are accomplishing, their group morale will likely improve. Another good tactic is to take along the first sergeant, the section NCOIC, or the operations officer on these visits. One Army battalion commander related he made it a point to often visit the troops with his sergeant major—when the battalion saw the top leadership together, they gained a sense of unit cohesion.

What about the relationship between the officer, enlisted, and civilian personnel in the squadron? Experienced commanders will relate that squadron relationships were one item to which they pay particular attention. All will agree there is little room for anything other than a strictly professional relationship among squadron members while on duty performing the unit's mission. This belief holds true regardless of the type of squadron or mission. This professionalism will build unit cohesion because it emphasizes the military nature of duty in the armed forces. Off-duty relationships will be discussed later.

> **PROVERB FOR COMMAND**
>
> "Mission identity = success."

To help build a sense of pride and mission accomplishment, a former C-141 squadron commander emphasized his unit's long history and military tradition. He wanted to ensure that his crewmembers realized they were not operating an airline, but rather a military operation that had a long history of service to the nation. In doing so, he felt he was able to build a sense of purpose in the unit's mission and help increase squadron pride.

Regardless of what methods are used to increase unit cohesiveness and morale while his personnel are on duty, if a commander doesn't also pay attention to off-duty activities, the battle will be lost and the mission will suffer.

Off-duty-related Examples

Off-duty activities can build unit cohesion and morale as well as duty-related activities. Squadron functions, parties, picnics, sports programs, and other activities all enable unit personnel to relate to each other in a relaxed atmosphere. These activities also give the commander a chance to relax with his or her people and see how unit members get along outside the workplace.

Ask 100 former squadron commanders to describe the off-duty relationships between unit members (of all ranks) and you may receive 100 different answers. There's nothing really wrong with this, each squadron is different, but as commander you need to know how your people relate to each other off duty and how these relationships affect the bottom line—the ability of your unit to perform the primary mission.

In conversations with commanders concerning relationships among unit members, a few constants continually appear. First, a squadron commander's own relationship with his unit personnel off duty will not be much different from what it is while on duty. Again a commander is not one of the guys, simple fact. Second, a commander needs to understand that many individuals will find it difficult to have an informal relationship with a more junior or senior squadron member while off duty, and then turn around the next moment and maintain a formal/professional relationship while on duty. A few people can do this with ease, others can't. Many commanders related how they quickly recognized this problem when they first took command. Some unit personnel were carrying their more informal off-duty relationships over to their duty relationships adversely affecting unit cohesion and morale.

Other commanders relate that their members were able to maintain a formal duty relationship while being slightly more informal off duty. The commanders here said that in these cases the unit benefitted from the relaxed off-duty atmosphere, became a very tight-knit outfit, and experienced no problems performing the mission.

A commander's job here is to know what works best for his or her squadron—there is no set answer that works best in all situations.

The object is to build unit morale and cohesion while ensuring mission accomplishment.

Squadron sports teams offer members the opportunity to participate together in an enjoyable activity in the name of the unit. Colonel Lawrence tells how his C-141 unit's sports program was a big boost for unit morale and allowed his enlisted and officer members to work together off duty.[7]

Some bases hold periodic "warrior" competitions where units vie for an overall base championship in a variety of competitions on one day. This type of activity can involve the entire squadron even if only a few actually compete. One former ATC commander tells about the "massive support team" his squadron set up for its triathlon competitors. They won more of these competitions than they lost, and the unit took pride in the accomplishment, knowing it was a true squadron effort.

Commander support for the sports program is also a good way to boost unit spirit. A commander doesn't have to attend every game, but periodic visits to team games will mean a lot to the competitors. If the commander enjoys playing the sport, he should play on the unit's team. In doing so, he shouldn't deprive other members of playing time, but if a commander was active in the base sports program before taking command, he shouldn't change because of his new position.

Squadron social functions come in all shapes and sizes. The type of functions one unit enjoys may not receive a favorable response by another unit. The following is a short list of ideas and thoughts on squadron functions that have been successful in the past:

1. Lt Col John Gibeau's large GLCM squadron emphasized flight social functions because when they deployed, they did so as flights.[8]
2. Col Bill Welser turned awards, decorations, and promotions into unit functions, where members could celebrate the hard work of their peers.[9]
3. Numerous commanders of relatively young squadrons relate that many of their parties were set up to easily allow those with babies to bring them along.

4. Periodically I'd have a squadron party in the squadron itself on a Friday evening. This was a central location for everyone, didn't burden anyone's home, made it easy for folks to come and go, was big enough to handle children if wanted, and enabled the families to feel more a part of the squadron.

5. Many commanders emphasized having parties that people enjoy and can all attend. Lt Col Don LaMontagne emphasized keeping party costs down so all his personnel could attend comfortably.[10] Innovation helps also—Lt Col Ron Love arranged for his squadron's children to taxi in a C-9 as part of their Christmas party.[11]

6. Lt Col Bob Lawrence managed to help build the morale of his troops in the dorm, teach them a new skill, and have a unit cookout all at the same time. He had a couple of squadron "experts" teach the troops the "ins and outs" of building a large brick barbecue pit. The pit was constructed adjacent to the squadron dorm, and it soon became a centerpiece for unit activities.[12]

7. Lt Col Craig Koontz commanded a large maintenance squadron in Germany. To help boost the morale of his young troops in the dorm, he arranged for a series of group activities with local German young adults. This proved to be a very popular activity; it enabled a large segment of his unit to benefit from the overseas assignment.[13]

8. Lt Col Steve Lorenz would host a "new-guy" party each month at his home where he invited all the new personnel who had arrived in his squadron in the last 30 days. This quickly bonded the new people together and to the unit, and it enabled them to meet their new commander in a relaxed atmosphere.[14]

How does a commander fit into a squadron party? However he wants is the best single answer, but here are a few thoughts and recommendations:

1. A commander always operates in a fishbowl, including at parties; its a good idea to get around to as many people as possible and then to leave before the party's over.

2. Some officers often have their spouses drive them home from a party because they may have consumed too much alcohol; a com-

mander should keep this practice to an absolute minimum, regardless of what he did before taking command—be smart, don't drink too much alcohol, lead by example.

3. If officers above the squadron in the chain of command attend a unit function, the commander has the responsibility to see to their welfare.

Unit cohesion and morale are important aspects of every commander's life; they present a big challenge to leadership and are absolutely key to squadron success. A commander who takes the time to understand how his people relate to each other will quickly figure out the best formula to meet the challenge.

Inspections

To the typical Air Force squadron commander, the terms ORI, ASET, MCI, CEVG, MSET, TACEVAL, UEI are not among the most pleasant acronyms in the English (read that Air Force) language. They all denote some type of formal inspection administered by higher headquarters personnel, usually at major command level. These inspections are "counters," in that "the big boys" assess your unit's ability to perform its mission and/or function as a military organization. They are not to be confused with local inspections performed by wing agencies or informal "staff assistance" visits, which, although important and useful, pale by comparison to the importance of a visit from the MAJCOM IG.

The bottom line here is that a formal inspection from headquarters is a combat situation! Your unit will need your guidance and leadership as the squadron commander, before, during, and after the inspection. Depending on how you conduct yourself and direct your unit's preparation, you can make the entire experience very easy and smooth or very difficult and rough for all concerned. You've seen many inspections come and go by the time you take one as a squadron commander—this time you get to do it your way!

The weeks preceding a big inspection are emotional times for the entire wing and base. People at all levels may have a tendency to "ping" and make decisions out of their normal character. A formal inspection is a "very big deal" for the wing commander—inspection results are usually seen as a good barometer of his leadership ability and potential for advancement. The Air Force historically selects a good portion of its brigadier generals from the ranks of its wing commanders. I've seen wing commanders "make it" and "break it" on inspections. As an inspection approaches, remember its meaning to the wing commander.

Inspections Preparation

How should you arrange your priorities for an inspection? Former commanders will all tell you to ensure *your mission is your number one priority!* The squadron may "look" great and "glitter" like gold, every "t" can be crossed and every "i" dotted, your leave program may be flawless and your weight control program perfect, but if your squadron cannot perform its primary mission, the IG team will have you for lunch. Any commander who strives to "look great" at the expense of his primary mission deserves to fail.

Most squadron commanders will go through at least a couple of inspections during their tenure and each will have his or her own way of preparing for the inspection team's "visit." Former commanders tell me that the type of inspection and the inspection philosophy of the major command are two important factors to consider when planning how to get the unit ready.

Lt Col Dave Knowles's philosophy was to perform the mission correctly every day using all unit resources as if you were going to war tomorrow. Because he didn't like special or exotic pre-inspection programs, he emphasized constant professionalism in the performance of his unit's B-52 mission.[15] Lt Col John Wilcox had a similar outlook. Flying a new, high cost system, his B-1B unit was constantly operating "in a fishbowl." They worked 100 percent at all times, and John felt this continually kept his unit ready for any inspection.[16]

One of the easiest ways to prepare for an inspection is simply to "read the mail," beginning with past squadron and wing inspection reports. Any problem areas identified in the past will definitely be closely scrutinized again. Your squadron should also receive the results of other wings' inspections by the same inspection team that is going to "visit" you. Pay attention to what these reports say and you will have a blueprint of what the inspection team is currently emphasizing.

Some commanders stress the need for a solid self-inspection program. One officer told me that he actively used a year-round program to constantly fine-tune his unit's mission and programs. Major inspections then just normally fell in line with his existing self-inspection efforts and erratic or extraordinary preparation efforts were not necessary. Lt Col Ray Chapman, 86th FTS/CC, used this type of approach and involved all unit personnel, emphasizing that the squadron would succeed or not succeed *together*.[17]

Other commanders relate that routine, year-round self-inspection programs are better in theory than in practice! They would appoint a small team of experienced supervisors to specifically run through the entire squadron self-inspection program a couple of months prior to the inspection in order to identify potential problem areas. Timing is crucial; self-inspections shouldn't wait until the last minute. They necessitate solid documentation and, above all else, require an honest, close look at squadron performance. Any IG team worth its salt will really "dig in" when they see a squadron's self-inspection results state that the unit has absolutely no discrepancies!

Here's a TIP—if your command emphasizes the self-inspection program, ensure your sections or flights are writing up every self-inspection discrepancy they find, no matter how trivial. Then, further ensure that a follow-up action correcting the action has been well documented. Go overboard to show the IG that you are totally committed to this program!

Another thought—be careful how you allocate resources when preparing for an inspection. Every time you apply additional resources to one squadron area, it can briefly cause a weakness in another area. As commander, you should keep a good grasp on how your resources are being used to ensure no area is neglected while other areas are receiving extra attention.

Most former commanders will state they prefer longer-range inspection preparation as opposed to a short, intense, panic-type plan. Lt Col Steve Roser, 14th MAS at Norton AFB, California, ran a building-block approach using his section chiefs to oversee each phase.[18] Lt Col Tom O'Riordan set up specific objectives for his B-1B squadron to meet at various times before the inspection. This method gives the commander certain benchmarks with which to judge how the unit is doing relative to the overall preparation plan.[19]

Lt Col Bill Looney, 22d TFS at Bitburg AB, Germany, accomplished a lot of precoordination with other base units and agencies that were closely associated with his squadron's mission. By working closely over the long haul with these other squadrons, Bill was able to help build a sense of teamwork in the wing that paid off before and during inspections.[20]

Lt Col Don LaMontagne, 492d TFS/CC at RAF Lakenheath, United Kingdom, prepared for an upcoming ORI by ensuring his flight crews were fully mission qualified in the aircraft, as opposed to merely being qualified "on paper." He personally handled the plans for the inspection while his operations officer worked the normal, day-to-day functions of the mission. Don continually stressed the need to hone the fighting skill of the squadron—focus on the mission![21]

A long-range preparation plan for your facility, building, or squadron area can also pay big dividends. I used this method to allow me to concentrate fully on my unit's primary training mission as we got closer to the inspection. The squadrons at Mather AFB do a lot of self-help work to improve their areas, and this can take a lot of time and resources. I set a deadline of 60 days prior to the inspection as the last time I wanted to see any major self-help work going on in the squadron. I wanted every bit of painting, paneling, carpeting, and repair work to be completed by this time, because I really wanted the focus on the mission close to the visit. Further, I also told my people that I would ask for only one Saturday of their time before the inspection, the last Saturday. That day we had a big work party, permanent party only, where everyone thoroughly cleaned his or her area, including me, solo in my office. I don't see any sense doing this too early and using this last weekend to bring everyone together creates a team atmosphere before the "combat." We ordered pizza for lunch and had a keg in the lounge. As we finished that afternoon, we cleaned the lounge last, rolled the keg out the back door and immediately locked up the squadron until Monday morning. The place had never looked better.

Most squadrons have an ample mixture of experienced and inexperienced personnel, chiefs and airmen, lieutenants and lieutenant colonels. A few unit personnel have never seen an inspection before, some have seen three or four, and a couple have just about "seen it all." Using experienced personnel in the squadron to literally teach the youth of the unit the ins and outs of a formal inspection during a commander's call is a useful tactic.

Lt Col John Bell relied heavily on his senior maintenance supervisors and his first sergeant to accomplish this task. These folks had seen many inspections, knew what to do and could teach the young troops the ropes. John appointed one of his senior NCOs to develop an inspection preparation plan. He initiated the overall process by meeting with the senior unit personnel to discuss weak areas and decide where to concentrate their efforts.[22]

Another commander knew he had two difficult problem areas in an otherwise smooth operation. He appointed one of his assistant operations officers to concentrate his efforts solely on these two items and detach himself from the rest of the preparation. Being able to devote himself totally to the two specific problems, he remedied the situations before the inspection.

LTC Steve Rippe was commander of a large Army battalion. He appointed one senior NCO to oversee and standardize each functional area (leave, NEO, security, and so on) that spanned his 900-man unit. One NCO was responsible for the leave programs of each company, one for the NEO programs—in short, one for each program. These experienced individuals not only helped each company commander prepare for the upcoming inspection, they simultaneously standardized many actions and programs in a very large organization.[23]

Personal confidence and coolness under pressure from a squadron commander will greatly help a unit prepare for an inspection. If you, as the boss, start pinging off walls and acting as if your hair is on fire, your people are likely to do the same and nothing will get accomplished. However, if you have a solid game plan and trust your people, your confidence will probably rub off on your unit and result in success. All members of a squadron want to do well on an inspection even if they despise these visits. With proper guidance, your people will prepare well if you show faith in their ability and desire—let them do their job!

An example: Air Force 35-10 standards are always a big item on Air Training Command inspections, especially for the students. I've seen some squadrons at my old base make a habit of conducting a full-blown open-ranks inspection before the IG team arrived. I always felt this was far too time-consuming and irritating to everyone; besides I just didn't want to micromanage that issue. Instead, I told my flight commanders what I expected for standards during the inspection and encouraged them to spend some extra time with the student classes. I trusted their ability to handle this facet of the inspection, and we had absolutely no problems during the visit.

Proverb for Command

"Tell what to do, not how to do it—then let them do it."

Before we leave the subject of preparing for an inspection, I want to touch on what I'll call "forced" preparation. Some squadron commanders relate how they received strict "guidance from above" about how to prepare for an upcoming inspection, including the exact times of the weekend they and their people were expected to be in their offices. In a couple of cases this was a wing-wide operation (remember the wing commander's stake in an inspection). Some advice here—it is sometimes hard to understand what your wing commander is driving towards, especially if he or she doesn't communicate well. The essential dilemma is balancing your efforts as a unit commander into the wing commander's vision. Sometimes it seems as if there is no vision. However, a squadron commander must be able to mold the squadron's operation to fit stated wing objectives, and he must get the squadron behind this effort.

During the Inspection

Once the inspection team finally arrives, you will probably be relieved to get the inspection underway. Your leadership will play a big part in your squadron's success during an inspection, and most any former commander will advise you to get out front and *literally* lead the way.

A flying squadron commander who takes the first checkride in an inspection on a tough mission will definitely lead the way. When you step back and think about this tactic, it makes good sense for

> **PROVERB FOR COMMAND**
>
> "Preparations for inspections or competitions can be overdone and actually get in the way of mission training; be careful here."

several reasons. First, your people will appreciate your close involvement in the inspection and putting yourself at some risk. Second, it will get your unit off on the right foot because you are most probably very experienced in the weapons system, and there are not many flight examiners who will "take out" a squadron commander unless there are some big problems during the flight. Finally, any inspection team will closely investigate a squadron whose commander tries to duck the schedule during an inspection.

A commander can also lead his unit during an inspection by working closely with the inspection team. Lt Col Buzz Moseley commanded the F-15 unit at the Fighter Weapons School at Nellis AFB. He prearranged his unit's flying schedule in such a way as to give the inspection team maximum opportunity and flexibility to fly its required number and type of sorties. He was sensitive to the fact that the inspectors had a job to do and figured that everyone would benefit by setting the flying schedule as easily as possible.[24]

Lt Col Bill Looney, former 22d TFS/CC, believes that you need to take some risks to really do well in an inspection. He had total confidence in his unit's expertise in air-to-air tactics and wanted to show the inspection team something besides the usual two aircraft versus two aircraft (2v2) scenario. During the inspection he scheduled some 8v4 flights and other more difficult scenarios that his squadron had practiced. Just launching 12 aircraft on a single mission is not an easy task, and the inspectors were impressed with the obvious emphasis on perfecting the unit's primary mission. Bill certainly took some risks here, because a lot can go wrong 8v4, but it proved a big success in the long run.[25]

Four other areas of inspection advice often mentioned to me by former unit commanders were:

- Be smart and put your best foot forward.
- Be positive, honest, and above board.
- Put your personal expertise to work.
- Stay informed during the inspection.

Some inspection teams want or need specific information either before they arrive or immediately when they get to your base. Be smart and provide them everything they want as soon as they'd like. If there are certain portions of this information you are particularly proud of and want to emphasize, make sure you highlight it. One commander described a new squadron program which had just been completed prior to the IG team's arrival. This was the first such program in the command and dealt directly with the unit's mission. He went out of his way to emphasize this new facet of his operation, and the inspection team quickly took to it. In fact, the team spent so much time looking at the new program, they didn't have time to scrutinize closely some of the other, possibly weaker, areas of the squadron.

Another facet of being smart concerns scheduling your people during the inspection. While you shouldn't tell your weaker people to go on leave for two weeks, don't schedule your newest personnel the toughest tasks when the IG team is in town. You wouldn't schedule like that in a war if you could avoid it—remember, this is a combat situation!

If the IG team finds some problems in your unit, don't spend a lot of time attempting to "talk your way out it" unless the team is actually wrong. For example: an inspector finds a couple of discrepancies in your leave program; acknowledge the errors, ask for suggestions on how to prevent such occurrences in the future and move on. Most commanders who have been through an inspection will tell you that being honest and above board is the best policy to follow. You can cut your losses by admitting a problem when found and pressing forward to other topics.

Proverb for Command

"Your gut feel about something is probably much better than what you give yourself credit for."

As the squadron commander, you will know your unit's mission and people better than anyone else, including the inspectors. Sometimes you may be the command expert in certain areas, and here is where you can really put your expertise to work. During one inspection I learned about, the IG team was coming up with some findings that just plain were not correct. They apparently based their opinions on a slightly different mission than that performed by this particular squadron. The squadron commander involved happened to be the most experienced individual in the entire command at this mission. This particular commander loved a good fight and felt strongly about the alleged discrepancies. So, he firmly stood his ground and almost "bullied" his way through the situation, finally convincing the IG team that they were wrong. When you know you're right, stand your ground and fight; your unit deserves your support.

Commanders also tell me that inspections can be "turned around" during the middle of the visit by merely keeping informed and in contact with the inspectors. A smart commander will ask for a periodic brief or "how goes it" from an inspection team. He will also keep his ear to the ground to pick up general trends around the wing and other squadrons. I've heard of cases where commanders learned early on that the IG team was hitting a certain area very hard and then had time to do last minute work on the area before the inspectors examined their unit. Another paperwork example: one particular inspector has already hit other units hard over a specific error in file procedures; if you find out about this early, you may have time to quickly correct any similar problems with your unit's files before the inspector visits your squadron.

After the Inspection

This is the easiest part, but an important part all the same. Postinspection—former commanders stress learning from the inspection and documenting what happened.

One technique I picked up was to hold a special squadron meeting solely to discuss what happened during the inspection. This type of internal squadron cross-feed enables unit personnel to see the bigger picture and to benefit from what the team may have suggested. Inspectors will often pass along valuable tips on operations and programs they've seen at other locations. More important, making this information available to your entire unit may help improve your operation.

Inspection documentation is considered vital by many former commanders. One officer tells me he kept a fairly detailed log of daily occurrences during the inspection and then wrote an overall summary after the inspection. These documents were then placed in a special squadron file for future use. You will probably not be in command the next time the particular type of inspection rolls around again, but you should do what you can to assist your successor.

Documenting what happened during an inspection can greatly help when it comes to writing future personnel performance reports. Your people work hard for an inspection and should reap the benefits of a job well done.

It's over—now back to business as usual!

Notes

1. Lt Col Bob Lawrence, interview with author, Ft. McNair, D.C., 9 November 1989.

2. Lt Col John Gibeau, interview with author, Ft. McNair, D.C., 28 November 1989.

3. Lt Col Steve Roser, interview with author, Ft. McNair, D.C., 31 October 1989.

4. Gibeau interview.

5. Lt Col Buzz Moseley, interview with author, Ft McNair, D.C., 17 October 1989.

6. Blackmore interview.

7. Lawrence interview.

8. Gibeau interview.

9. Col Bill Welser, interview with author, Pentagon, 27 October 1989.

10. Lt Col Don LaMontagne, interview with author, Ft McNair, D.C., 5 October 1989.

11. Lt Col Ron Love, interview with author, Ft McNair, D.C., 1 November 1989.

12. Lawrence interview.

13. Lt Col Craig Koontz, interview with author, Ft McNair, D.C., 29 November 1989.

14. Lt Col Craig Koontz, interview with author, Ft McNair, D.C., 16 October 1989.

15. Lt Col Dave Knowles, interview with author, Ft McNair, D.C., 15 November 1989.

16. Lt Col John Wilcox, interview with author, Ft McNair, D.C., 7 November 1989.

17. Lt Col Ray Chapman, interview with author, Ft McNair, D.C., 19 September 1989.

18. Roser interview.

19. Lt Col Tom O'Riordan, interview with author, Ft McNair, D.C., 10 October 1989.

20. Lt Col Bill Looney, interview with author, Ft McNair, D.C., 24 October 1989.

21. LaMontagne interview.

22. Lt Col John Bell, interview with author, Ft McNair, D.C., 11 October 1989.

23. LTC Steve Rippe, interview with author, Ft McNair, D.C., 2 November 1989.

24. Moseley interview.

25. Looney interview.

Chapter 3

People—The Key Ingredient

A Soviet general and an American general were discussing the Kremlin and Pentagon buildings one day. The Soviet general noted that the Kremlin structure had four sides, corresponding to the four pillars of a strong military. He asked his American counterpart why we had mistakenly constructed a building with five sides instead of four sides. The American general thought for a few moments . . . let's see, we have force structure, strategic modernization, sustainability, readiness and, oh yes . . . and then replied, "The fifth side of our Pentagon is our people—advantage blue!"

As we proceed through the 1990s, the United States Air Force is blessed with a highly qualified and motivated force. We have never been healthier in terms of the quality of our people. If you have been on active duty for more than 15 years now, ask yourself if you could even be accepted for active duty today. Chances are, your answer is, "No, not unless I got a personal recommendation from the president!"

People are what make the Air Force tick—they are the most important ingredient of any squadron. "Take care of your people, and they will take care of the mission" was the most prevalent concept I've heard from commanders. Mission and people should be the top two items on a squadron commander's agenda.

A good commander works his people, works with his people, and works for his people every day. Personnel issues are many and highly diversified; a skilled commander will make every effort to stay on top and often ahead of these issues. I'll initially develop this topic of people by examining a commander's (and a squadron's) relations and interactions as people move through the unit. This "life cycle" approach will start with a new member's arriving on station and will progress through the individual's advancement in the squadron, career development, promotion, reassignment, and finally departure. I'll then address four other specific people topics: awards and decorations, discipline and UCMJ matters, crisis situations, and the role of spouses and families.

The "Life Cycle"

"Taking care" of unit personnel is not solely a commander's job; it is a squadron's job. I like the way Lt Col Craig Koontz challenged his unit during his first commander's call: "If we don't take care of ourselves, nobody else is going to. We may not always be one big happy family, but we are always going to be a family." We start at the beginning.[1]

Welcome

By far the best welcome program I've run across was described to me by Lt Col Bill Looney, who commanded the 22d Tactical Fighter Squadron in Bitburg, Germany. He felt a solid welcome program was absolutely vital at an overseas location, and he structured his program to bring the new unit member and his family into the squadron fold quickly.[2]

The 22d would normally receive initial notification of a new member a few months before the actual arrival. The new member was assigned a sponsor, who then sent a packet of information and letters about the squadron, mission, base, and the local area. The packet was closely followed by a personal phone call to the new member, wherever he was in the world. This phone contact was always made, regardless of the difficulty. Once the sponsor made contact, he gathered as much information about the new member and his family as possible. This included exactly where and when they would arrive in country, plans on housing, needs for schooling, if they would like coffee mugs, T-shirts, patches, and so on, and if they needed advance work accomplished for any special problems or circumstances.

The sponsor would then immediately order all the patches, name tags, coffee mugs, and T-shirts. He paid for them out of unit funds, to be reimbursed later by the new member. The sponsor also investigated the possibilities for meeting housing and transportation needs so that this information would be available when the new member arrived. The squadron commander booked the Transient Lodging Facility (TLF) for the maximum number of days and the squadron always got the room a day early. They then proceeded to stock the TLF with food, drink, and any critical items such as diapers and baby formula. This precluded the possibility of the new family's suddenly finding themselves without essential items at midnight in the middle of Germany.

The sponsor always met the new member and family at the airport and drove them to the base, so the first person they met in this strange land was someone they had talked to before. When they got

PROVERB FOR COMMAND

"Sponsorship for new people is essential."

to the TLF they were already checked in, the facility was clean, and it was the right temperature (air conditioning or heat already on if necessary.) The name tags, patches, coffee mugs, T-shirts, and other items were laid out on the bed for them, and they had plenty of food, drink, and supplies for a few days. They did not have to spend an hour checking in at the billeting office; they did not have to search out an unfamiliar base and countryside for food and supplies; they did not have to hassle with anything. They did have to relax—and they could, because the squadron had taken care of everything for them.

An Air Force member who receives this type of treatment upon arrival at a new unit will most likely do everything he or she can in the next few years to pay the squadron back for their hard work. A welcome program such as the one described here instantly bonds the new member and his family to the unit and can't do anything other than enhance mission effectiveness. Perhaps this program is one reason the 22d TFS often won wing and command awards for excellence.

In-briefings

After getting initially settled in at a new base, the new member next needs to get acquainted with the mission, the people, and the organization of his unit. The amount of detail that he needs to know will depend upon exactly what his job will entail. Generally, but not always, more senior officers, NCOs, and civilians will need to see a bigger picture than a junior officer, airman, or civilian.

In most all squadrons the commander conducts an initial interview or in-briefing with a new member. This is usually an informal, but still very important session. While this first session allows the commander to get to know the individual better, its main purpose is usually to let the commander explain how he sees the squadron and the mission from his position on top. Commanders should take this first talk very seriously, making a good effort to prepare properly what thoughts they want to get across to the new member. Sometimes, in a very large unit, the commander may brief more than one individual at a time. Regardless, those commanders who conducted such group briefings still took the time to prepare fully.

Besides talking to the commander, a newly arrived member will also need to meet with other key unit personnel. These key people will explain their facet of the squadron's operation and organization, how they fit into the overall unit picture, and how the new member will interact with them. From a new member's perspective, the key people in the squadron are most likely those in his own supervisory chain of command plus a few others, most notably the first sergeant. Once our newly arrived unit member has been in-briefed, he's ready to begin his duties or to begin a training program to qualify him for his duties.

"Key Personnel"

Before progressing further, I want to expand on this concept of key unit personnel, this time from the perspective of the squadron commander. Just who are a commander's key people and why are they key?

One commander explained that he defined key in this instance to mean someone to whom he gave important authority for the running of one aspect of the squadron. He also added these individuals were key because they did an excellent job and allowed him to "command instead of micromanage" the squadron. This officer commanded a large maintenance squadron and listed his first sergeant, supervisor, superintendent, two branch chiefs, and his secretary. He did not list a couple of his other branch chiefs as key because they had to be closely supervised themselves.

Most commanders will explain that getting to know who your key people are and how they help you command is one of the first tasks a new commander needs to accomplish. Once a new commander has a firm grasp on the strengths and weaknesses of his personnel, he then knows where and when he can delegate authority (not responsibility) to run the unit efficiently.

As in the example above, commanders of large units place a premium on having a top-notch first sergeant. As one lieutenant colonel put it, "There is a special place in heaven for first sergeants." A good first shirt will always be in the know about what's really happening in the unit at the grass-roots level. Before they earn their diamond, they receive special training in their duties and are prepared and eager to serve as the commander's right hand. By all means, if you have a good first sergeant, let him do his job!

If the first sergeant is the right hand, then the operations officer (or supervisor/superintendent) is the left hand. In most units, this individual normally runs the daily operations/mission of the squadron. Again, if you have a good ops officer, let him do his job. For some commanders this is not always easy to do. Although few commanders were ever first sergeants, many commanders were ops officers at one time or another, most likely enjoyed the job, and consider themselves subject matter experts. A sharp commander will make an extra effort to allow his ops officer the latitude to run the operation as he sees fit, maybe guiding him a little here and there, but dictating only when absolutely necessary.

Individuals who supervise large numbers of squadron personnel, such as branch chiefs, section heads, superintendents, and flight commanders, are also key to the smooth accomplishment of the unit's mission. These key people are not always military. Lt Col Kathy LaSauce, commander of the 93d Aerial Port Squadron at Andrews AFB, relates that her civilian supervisor in the passenger terminal (many, many DVs) is absolutely critical to her mission. She told me this individual has such a fantastic "corporate knowledge" of passenger operations that she totally trusts him to run that highly visible portion of the squadron.[3]

> **PROVERBS FOR COMMAND**
>
> "The first sergeant is vital, you must determine his or her effectiveness early on."
>
> "Help the Air Force bring the young people up right—work hard to develop their skills."
>
> "Find out who you can trust and then trust them to do the job."
>
> "Use your immediate supervisors and make them do their job."
>
> "Give people the tools to do their job."

These supervisors often require good people skills and a certain amount of leadership. They are also the individuals who often oversee the training of our newly arrived squadron member—they mold the future of the unit. For those reasons commanders want top quality folks in these positions, but may not always get what they want.

Later on in this chapter I'll discuss what happens when the commander is unhappy with his squadron supervisors. First though, let's go back to our new squadron member as he progresses through his assignment in the unit.

Counseling

Webster's dictionary defines counseling as: 1. to give advice, 2. to recommend (an action, etc.). Counseling takes on many forms in the Air Force, and a squadron commander will find himself involved in most forms during his tenure at the top. Counseling is not easy—it takes a certain amount of skill and preparation to effect a productive counseling session.

One type of Air Force counseling session prevalent today is the performance feedback session. Every single commander I talked with felt that the inclusion of performance feedback in the recent changes to both the enlisted and officer personnel systems was the best personnel step taken in years. The only criticism I've heard relative to the new performance feedback mechanism is that it should be mandatory for all ranks.

A squadron commander's feedback sessions should never be "square fillers"—he or she needs to have a firm grasp on giving honest, fair, and comprehensive feedback for several reasons. First, individuals who work for the commander, and thus will receive feedback directly from the commander, are normally the senior supervisors in the unit. Their performance largely influences the success or failure of the squadron. A commander who treats feedback sessions lightly not only cheats the recipient out of useful advice but also cheats the squadron as a whole.

Second, for you as a commander, these sessions are particularly useful tools with which to guide your senior supervisors in the direction you wish the unit to follow. You have a periodic counseling mechanism already built into the system—you don't have to give the impression that an individual has done something wrong to get called in for counseling. The people that work directly for you will be more receptive to your counseling (good and not so good points) because of the system. Finally, you are the commander and operate in a fishbowl, remember! One commander related how within two days of giving any performance feedback session, most of the squadron knew what had been discussed and "how the session went." Squadron members look up to the commander and expect a professional effort in whatever he does—when he doesn't put forth a good effort they will know and may respond in kind.

A commander can influence any counseling session by the physical setting in his office. In the situation of our squadron member who is getting performance feedback, most commanders I talked with recommended an informal atmosphere. They suggested that the commander and the member sit down at a table, in two comfortable chairs, or on a large couch. This type of session should be two-way, a give and take of ideas and thoughts, with the commander leading the way. Most commanders did not have the member formally "report in" for this type of session.

Commanders are also heavily involved in career counseling. As a minimum, squadron commanders should periodically career counsel those individuals they directly supervise. Many commanders couple performance feedback and career counseling into one session. Commanders relate that they prefer to keep this type of counseling on an informal basis also. They also say that career counseling requires a good deal of commander preparation.

One former commander explained how he kept a large notebook by his desk with a wealth of career information on every individual he supervised—complete with a career brief and notes on past performance reports. He would thoroughly review this information before talking to the member and would focus the discussion on development/advancement in the squadron and wing, career goals/desires after this assignment, and the individual's promotion potential. Each section of the discussion was a two-way conversation designed to allow the member to express his goals and desires and allow the commander to offer solid guidance and advice.

Lt Col Bill Welser commanded a C-141 squadron at McGuire AFB. He explained how he always set special times for career counseling sessions to allow the member's spouse to attend. He felt strongly that today's Air Force members do not make career decisions on their own, but will almost always fully involve their family.

Once, Colonel Welser had a young pilot who was considering separating from the service. The captain talked about dissatisfaction with some aspects of the Air Force and how the airlines looked so attractive. Bill did not try to "hard sell" the pilot on the Air Force, but presented all sides and issues of the decision for consideration. Then, Colonel Welser asked the pilot's wife how she viewed the situation. She related that she had a solid, career job with a law firm in Philadelphia, and did not wish to leave the area under any circumstance. Suddenly, the picture had clarified—now Bill knew the driving reason behind the officer's desire to leave the service. Again, Colonel Welser discussed all aspects of this decision with the couple and asked them to thoroughly think out the pros and cons of each side before making a final decision. The captain did separate from

PROVERBS FOR COMMAND

"Know your people."

"Listen to your people and genuinely care about them—be positive and don't micromanage."

"Understand that a big part of your job is to develop the next generation of leaders."

"Help people when necessary, don't just criticize them or let them stumble."

the Air Force, but at least he had a commander who cared and helped make the decision an educated one.[4]

Other types of commander-initiated counseling sessions are not totally pleasant. Often a squadron commander needs to counsel a member about a discipline matter or a definite lack of performance. This form of counseling also requires preparation by the commander, but the overall style is normally more formal. One commander told me if he were counseling an individual for the first time for lack of performance, he would couch his words around the idea that the individual was letting the entire squadron down by not doing his job. If the commander had to counsel the same individual more than once about the same subject, the counseling session often turned into a "one-way conversation." In a one-way conversation, one person talks and the other person listens and says, "yes sir" a lot. Also, one-way conversations are usually characterized by the listener's "reporting in" and standing . . . and standing.

Hopefully, our squadron member who arrived in the unit at the beginning of this chapter has only seen positive counseling sessions with the commander. He has long since been fully trained in his specialty (the ops officer, supervisor, or superintendent has done his or her job well) and is now eager to assume new responsibilities in the unit or wing.

People Decisions

Some of the most important commander decisions involve placing the right person in the right position in the squadron. This is another way in which a commander molds his or her unit and focuses the squadron in the direction he or she desires. In general, there are three basic categories of members in an Air Force unit:

1. Most officer and enlisted personnel are "solid citizens." They consistently do a good job, cause few problems, and will progress through the squadron and Air Force at a good, steady pace as they rise towards their potential.
2. A few members will definitely be out ahead of the ballgame regardless of the task or duty. They tend to be fairly innovative, very energetic, and extremely flexible. A sharp commander will be able to pick this type of individual out of the crowd within two months of the person's arriving in the unit.
3. A very few members reach the limit of their potential early. They may fail at certain duties and tasks or even cause the unit and the commander problems. Most often, these people are also easy to identify early.

The first type of person mentioned here is the easiest for a commander to place in the unit. The other two types are more difficult.

Let's say that our squadron member who came into the unit months ago is a solid citizen. This individual has done everything asked of him and prior career counselings indicate that he is ready and eager for added responsibility.

Most commanders will tell you that at this point one of two things needs to occur. If the commander has a squadron position available, and this member is the best qualified, the move should be made. If there is not an available position or the individual is not the best qualified then the commander or the immediate supervisor should explain to the member why he is not being moved at this time. I would recommend the commander be the one to talk to the individual—this action will let the individual know the "big boss" cares and is aware the member is ready or not ready for advancement.

Next scenario—our member has been with the squadron three years and is one of the very best midlevel supervisors, a job he has held for the last nine months. He is definitely in the "well-above-average" group, and you, as commander, have no higher position in the squadron where you could move him. However, there is a key wing-level position opening up next month for which the person is qualified. At the same time, you realize that you are due a major IG inspection in five months. OK commander, your decision!

If faced with this situation, experience recommends you nominate your individual for the wing job and fight like hell to ensure he gets it. Lt Col Steve Lorenz related that he worked hard to get his people wing jobs and also to later bring them back to the unit in even better jobs. Steve believed this movement helped the squadron and the individual.[5]

Another commander told me he tried to "influence" the direction of the wing by constantly placing his people into key wing positions. At one point in time, he said that 55 percent of the wing jobs were held by people who had come from his unit and there were five similar squadrons in the wing! Additionally, all of the wing-level executive officers were out of his squadron—he never got "blindsided" by anything coming his way from above. This commander also said this approach allowed for a healthy turnover of supervisory positions inside the squadron. Realizing this situation, his people naturally put out the extra effort, knowing they would be recognized and the boss would take care of them. The commander also had confidence his people could be thrown into a tough, new job and always land feet first.

Final scenario: one of your midlevel supervisors (branch chief/flight commander) just isn't hacking the job. He had been a solid citizen before as a worker but had apparently reached his potential in that duty. He is a fairly popular individual in the unit and you, as commander, have previously counseled him on his shortcomings in his supervisory performance. Again, the ball is in your court!

All of the former commanders I talked with said that this situation requires the commander to replace the supervisor. This same group of commanders also said that firing a squadron member is one of the toughest things to do. In fact, one of the final questions I asked each commander during my interviews was, "If you had it to do over again, what would you do differently?" About 35 percent of the commanders replied they should have fired a certain individual but didn't, for a variety of reasons.

PROVERBS FOR COMMAND

"Push your people for wing jobs—develop leaders."

"Take care of your people and your people will take care of the mission."

"Find out who your future leaders are and train them."

"Let people be part of the team, let them spread their wings."

All of the commanders who did fire someone emphasized the need to counsel the individuals first on any problem areas and give them the chance to improve their performance. During this counseling, the commander should explain exactly what is needed to remedy the situation: that is, what you, as commander, expect.

Lt Col Dave Morrow had to fire one of his F-15 flight commanders. Dave explained that the action was a result of a series of incidents and, before taking action, he made sure he had the complete information on each occurrence, including the flight commander's side of the story. Once he decided he had to fire the officer, Dave called the flight commander in for a meeting and told him straight out that he was being replaced. The next day, Colonel Morrow explained his action to the rest of the squadron—this squelched the rumor mill and told everyone exactly what the commander expected out of a squadron supervisor.[6]

TIP—a commander should always back-brief his boss after firing one of his unit supervisors, officer or enlisted.

Promotions

Let's again go back to our squadron member. He's been in the unit three years now and is up for promotion next cycle. What do you, the commander, have to do with this?

The squadron commander has a great deal to do with promotions in his unit. The commander influences the duty and level of responsibility of each squadron member, the type of performance report everyone receives, and often the amount of effort each member exerts towards getting promoted.

Before beginning to influence promotions, a good commander will ensure he has a basic working knowledge of the overall subject of promotions. This is not always as easy as it may seem—consider a large squadron that includes civilians from clerk-typist to section supervisor, enlisted personnel from airman to senior master sergeant, and officers from second lieutenant through major. That's three promotion/advancement systems with subsystems imbedded in each! As a new commander, if you feel uncomfortable about your knowledge in this arena, go to the CBPO and ask questions—remember, there are no dumb questions!

Experienced commanders tell me the next step is to ensure their people are knowledgeable about their own personal promotion situations. Just because every member of a unit is a "person" does not mean he or she is a "personnel" expert! Many commanders include the subject of promotions during career and performance counselings. If the squadron is so large that it precludes the commander from counseling each member, the supervisors who do career counsel members then need to discuss promotions during these sessions.

What type of "promotion information" do you as commander (or a supervisor under you) need to know about each individual and to ensure the individual is aware of also? Here is a short list for starters:

1. The timing of the next selection process for which the individual is eligible. Amazingly, some people are unaware of the dates or have heard and believe an errant rumor.

2. If there is testing involved in the selection process, when the tests are being administered. I've seen promotions lost because individuals missed their testing dates! A sharp commander will note this on some type of unit calendar and make sure his people are present for testing. One commander also related he would make every effort to give his people some study time during the week before the testing.

3. The status of the individual's record. The Air Force military promotion process does not allow for personal interviews; selection boards have to rely on written records that portray a great deal of information. This record needs to be complete and up to date! Commanders recommend first looking at the CBPO's copy of the record along with a computer-generated brief. This will uncover the existence of most any problem. However, the record that counts is located in San Antonio, Texas—if the individual can't get to Randolph AFB to review the record personally, he or she should write for a microfiche copy to review.

4. If another performance report on top of an individual's record would help the member get promoted (or not get promoted if you see it that way), then make it happen.

5. Photos, re-occurring forms, and other documentation all need to be updated as necessary. TIP on photos—never have anyone go to the photo lab alone for an official photo, always have another

squadron member go along to eyeball the pose, uniform, and individual, from a slightly different angle and perspective than that of the photographer. This will reduce the number of retakes!

6. The Air Force has not told people to "blow off" PME, they have told them to take it at the appropriate time. The same thing holds true for civilian education. The Air Force's emphasis on the mission and away from "square fillers" is right on target in my book, and way overdue. However, the young troop who thinks PME and education are no longer "players" will realize later in his or her career that a big mistake was made.

After all is said and done, remember that a commander can guide his or her people to the promotion trough, but those people have to drink from it themselves through job performance and proven potential for advancement. This holds true regardless of the type of promotion system and regardless of the grade involved.

> **Proverb for Command**
>
> "Be involved with your people."

Back to our trusty squadron member, who now has been selected for promotion and is nearing four years on station. MPC says its time for reassignment. Commander, what now?

Assignments

For openers, if you haven't done so already, you need to sit down with the individual and review four questions:

1. Do the assignment folks at MPC or MAJCOM have the latest information concerning the individual's desires? If not, this needs to be accomplished ASAP (as in NOW). Also, make sure that you are fully aware of these desires.

2. Is there any possibility of quickly working an action through a commander's involvement program? Some commands allow a squadron commander to directly work assignment actions. Commanders who have been involved in this type of program think its a tremendous improvement in the system. If this is possible, pursue it with vigor!

3. Has the individual contacted the agency that will initiate the assignment action? If not, he needs to do so and find out his current status.

4. Are the individual's desires realistic? If not, you need to be honest and tell him your opinion and possibly suggest better options for a career path.

After answering these four basic questions, the next step will vary, depending on the exact circumstances. The Air Force policy and trend is toward greater commander involvement in reassignment actions—so now is the time to get involved. While there is no set formula for getting involved in this process, here are a few commander TIPS as they were recounted to me:

1. Personally contact the people making the assignment and let them know of your interest.

2. Work closely with the individual involved. Tell him of your conversations with the assignment folks and ask him to inform you of his latest conversations.

3. If your squadron member is being considered or nominated for a specific position for which you feel he is well qualified, call the commander or supervisor who is doing the hiring and tell him why your person is best qualified. Inform the assignment officer you are making this call.

4. Sometimes you may have to work as a coordinator between two or more assignment agencies, such as MPC and MAJCOM(s).

5. Once a good assignment possibility develops, don't let the ball drop at the last minute. Follow through until the paperwork is in your member's hand.

6. Be prepared for some disappointments along the way—flexibility is the key to air power! The needs of the Air Force will, and

should, come first. Your involvement will, however, increase the chances the individual's desires will also be met.

Continuing on, our squadron member now has PCS orders. He's had a successful tour—has a promotion line number, has a good assignment awaiting him at his next base, and has been a very productive worker and supervisor.

Departures

This final segment in the squadron life of our member will also vary according to circumstance, and there is no set answer as to how you, the commander, should work the departure. Again, here are a few thoughts and ideas you may want to consider:

1. Once you know when an individual is leaving, begin planning for his replacement if applicable. If the departee is a supervisor, plan for adequate overlap or job turnover. This may require a series of internal moves that may test your ingenuity.
2. Do everything possible to give the individual as much time as realistically possible to take leave (if desired) and outprocess smoothly. Don't "abuse" him until the last minute!
3. Stay well ahead of the game on performance reports and end-of-tour/decoration-nomination packages.
4. Ensure he and his family are properly bid farewell by the squadron.
5. Always keep in the back of your mind the things you liked and didn't like when you departed various stations during your career. Let these experiences guide your command actions.

Congratulations—you have now successfully guided someone completely through a tour in your squadron!

The Good—People Recognition

Actively recognizing the accomplishments of squadron members is perhaps the best way for a commander to build both individual and unit morale. Napoléon once stated that "for a few yards of ribbon (medals), I could conquer the world." Colonel Lorenz once wrote in one of his squadron newsletters: "Ego is the number one facilitator."[7] Everyone has an ego, just as everyone likes praise and positive strokes. A smart commander will use these natural human tendencies to his advantage and establish a strong unit recognition program.

Such a program will require a great deal of work. There is often a lot of paperwork and coordination involved in recognition programs, all of which consume considerable time. Is it worth the effort, the hassle? Emphatically yes! Show me a squadron with a strong recognition program, and I'll show you a unit that excels in mission accomplishment.

In this section I want to break recognition programs into four categories: Air Force individual decorations (ribbons and medals), command and higher awards, wing awards, and squadron recognition programs.

Air Force Decorations

In the long run, Air Force decorations are the most important because they stay with the individual throughout and even past his or her Air Force career. Ribbons and medals are also evident every time the member puts on the service uniform. They are a continual source of positive strokes. For a squadron commander, there are two distinct facets of Air Force decorations: the actual awarding of the medal or ribbon, and the paperwork documentation that recommends the award.

Of the two facets, the actual awarding of the decoration is obviously the more appealing. This does not mean that an awards ceremony should be taken lightly or accomplished haphazardly. A well-done, professional awards ceremony can add to the positive impact of the award itself.

Most squadrons hold these ceremonies in conjunction with a large unit gathering, usually at the beginning of a commander's call. Often the member receiving the award is new to the squadron and the award is for work at the last assignment. Commanders tell me this situation presents a good opportunity to reinforce the unit's welcome to the individual. Numerous squadrons ensure the recipient's family can attend the ceremony. Some units traditionally give flowers to the spouse and hold a small reception after the event. The more a commander can personalize this event, the better.

Most commanders preferred to present these awards themselves, although a few often asked the wing or vice wing commander to present the decoration. This is a matter of personal preference.

TIP—a new squadron commander may want to check around first to ascertain if there are any base-wide standard award procedures. More than one new unit commander has made the mistake of not inviting the "right people" to an awards ceremony. Let's go back to the new squadron member for a moment. Individuals arriving at your unit from an assignment other than a school or training course will normally receive some type of end-of-tour decoration. In most cases, the Air Force's paperwork is completed and approved after the member reports to the new unit. As the individual's new commander, you should be on the lookout for this decoration. If one doesn't appear in a few months, find out why.

A former ATC squadron commander tells how he discovered two of his newer people missing decorations from their last tours. Both had arrived from the same overseas base. After a short investigation, the commander discovered three more squadron members in the same situation, all having PCSd from the same location. It turned out this overseas base (and especially a couple of squadrons there) was somewhat notorious for "dropping the ball" on end-of-tour decorations—once out of sight, out of mind. The commander tried to recoup what he could. After many overseas and stateside telephone calls and some coordination, he was able to "save" three of the five individuals and get their rightfully earned decorations approved. Many young captains, lieutenants, sergeants, and airmen will not speak up when they are a victim of this type of situation. As a commander, you need to stay on top of things and do what you can to help your people.

Decorations awarded at squadron functions need not be limited to new arrivals; in certain circumstances and with some extra work, awards can be approved and presented while the individual is still with the unit.

Lt Col Ron Love commanded the 11th Aeromedical Airlift Squadron at Scott AFB, Illinois. Many of the pilots in his squadron arrived straight out of undergraduate pilot training and had never received any type of medal. Ron would look at these and other individuals as they reached the three-year point in his squadron and, if deserving, would recommend them for a decoration. This process always took extra time, but he always had the award approved and presented before the member was reassigned. Colonel Love also had an ulterior motive here—he knew that this decoration would make his people more competitive in any type of selection process, so instead of waiting until the member was ready to PCS, he initiated the process as soon as possible. Moreover, by being proactive, he determined the suspenses, which allowed him to smooth flow the award workload.[8]

"Impact" medals are a good method of quickly rewarding superior accomplishments. An impact medal is one usually given for a specific achievement and is awarded soon after completion of the act or project. An Air Force achievement medal is excellent in this circumstance and, when used with discretion, greatly improves individual and unit morale. Currently, the wing commander is the approval authority for an achievement medal and the paperwork is not overburdensome.

Unfortunately, the Air Force has not followed in the Army's footsteps in lowering the approval authority on decorations. An Army battalion or squadron commander can approve an achievement medal, a brigade commander can approve a commendation medal, and only a meritorious service medal or higher has to go off-post for approval. Battalion commanders tell me their approval authority for the achievement medal adds to their command stature and their capability to lead troops. Hopefully, the Air Force will soon follow suit.

Lt Col John Gibeau commanded a ground launch cruise missile squadron in Belgium. He related how he instituted award ceremonies for good conduct medals. Many times these medals are not presented at any ceremony. John believes the Air Force Good Conduct Medal represents solid service to the nation and should not be taken lightly—as a commander he followed through on his beliefs.[9]

How about the longevity ribbon and its oak leaf clusters? I've seen these awarded at ceremonies also. The bottom line: any military decoration gives the commander a chance to publicly recognize the accomplishments of unit personnel—always a good opportunity!

The second facet of military decorations is the less appealing—the paperwork documentation recommending the award. There is not a commander alive who has not been frustrated at one time or another over the tedious process of putting together an award package. One former commander related how he tried for two years to get end-of-tour medals approved and awarded before the recipient went PCS. He failed every time because he ran into more roadblocks than can be imagined, both on and off base. He never gave up trying, but he said that this was one of his biggest disappointments as a commander.

But wait—as a squadron commander, before you begin the paperwork, you need to decide which award each individual should be recommended for, if any. Former commanders all recommend being honest, discriminating, and consistent. Commanders who violate any of these three principles run the risk of making the whole process unfair and meaningless. This isn't easy—often "guidance" will come from above. Squadron commanders should fight for what they believe is fair and just. They should also realize that they may not win every battle.

I lost more than one of these battles. At one point in time, those above me would not approve many medals for people separating from the Air Force. I recommended one of my separating officers for a Meritorious Service Medal because I felt the individual had earned it and it was consistent with other decorations given for similar performance. I did not feel the fact that this person was separating from the service entered into the equation. I had remembered a statement from a general officer I had worked for in the Pentagon years back who said, "Too often the military forgets that an individual who separates after 6-10 years of service has given the nation the best 6-10 years of his or her life!" In this particular case, I barely got a commendation medal approved for a really top-notch officer. I lost the battle, but I didn't give up the war.

Commanders tell me they normally inform an individual about a decoration recommendation. This includes telling people who have exhibited poor performance they are not being recommended for a medal and the reasons for the nonsubmittal. This may not be a pleasant task, but it comes with the command turf. Also, as in my case, if you find out a recommendation has not been approved or has been "downgraded," you need to inform the member of this fact quickly, especially if he or she is about to PCS.

In my opinion, the paperwork recommending someone for a decoration is second in importance only to enlisted and officer performance reports. Decoration documentation has to be precise and absolutely correct every single time, no exceptions. The various formats are spelled out in Air Force regs and are no secret to anyone. TIP—set up a good system in the squadron to work decoration paperwork well ahead of suspense dates. There is nothing worse than setting your hair on fire to complete a recommendation package at the last minute. It never works out right and always comes back to you for reaccomplishment! Many commanders appoint someone to oversee this process. Many unit executive officers and secretaries keep elaborate charts, boards, and suspense files on this item. Having an efficient system will pay huge dividends. I'll discuss this paperwork further in another chapter.

After you've submitted the paperwork, follow through with the process. Don't end up like the people on the overseas base I mentioned before and let the ball drop on decorations for members who have been reassigned—don't be satisfied until you've learned the medal has been approved.

Command and Higher Awards

Awards and recognition programs at the major command, Air Force, or higher level are normally very competitive. As a squadron commander, you will periodically receive notices (normally through the CBPO) inviting you to submit a nominee for various types of awards. These can range from command technician of the year to very prestigious national or international awards sponsored by organizations such as the Air Force Association.

The nomination package for these types of awards will often require a lot of work. The first question you will ask yourself is, "Should I bother?" Most commanders will answer, "Yes, if I've got somebody who fits the requirements and description for the award." Support your people whenever you can.

Before answering this question, a commander first needs to properly research the award to determine the exact selection criteria. Many times, the call for nomination letters will not contain a lot of specific information about the award. Command or Air Force awards are normally covered in some type of regulation or manual.

After determining the selection criteria, the commander then needs to decide who, if anyone, to nominate. The commander can either do this himself or he can delegate the task. More often than not, there will be someone in the unit who is suitable for nomination, even for many of the higher level awards. A squadron that continually submits "negative replies" for these types of awards is just cheating its people. All nomination packages are forwarded up the wing chain of command and often are reviewed by the wing commander. Anytime you can put the accomplishments of your people before the wing commander, do so—you are helping both the individual and the unit.

Some awards may honestly not fit anyone in the squadron. In this case, the commander should not submit a nomination. However, many commanders will tell you they were directed from above to submit a nomination—all complied! In this situation just do the best you can.

A former SAC squadron commander told how his unit established a system for all award nominations that would go off base. Over the years, the unit had built a fairly detailed set of files relating to off-base awards. The files contained the past five year's worth of nomination packages, applicable regulations or manuals, and all available information concerning past award results. Every three months, the commander, executive officer, and awards/decorations officer sat down and reviewed the upcoming quarter's award nomination requirements. They stayed well ahead of the ballgame to ensure timely, accurate, and quality nomination packages. They won more than their fair share of the awards.

Another former commander explained how his squadron used a similar type of system to assist in the annual preparation of their Air Force Outstanding Unit Award (AFOUA) nomination package. Along with keeping past unit nomination packages, the squadron secretary created a special file each year containing unit accomplishments. This readily available information was invaluable in preparing the yearly AFOUA package. The squadron had an impressive record in being selected for this award—an award that stays with each unit member throughout his career. Additionally, the squadron always communicated this selection to former members who had moved on to new assignments but had been in the unit during the period covered by the award.

Wing Awards

Most Air Force wings have similar awards/recognition programs. They are usually periodic in nature (often on a quarterly basis), competitive, and well documented in wing regulations. These awards provide a commander an excellent avenue to "showcase" the accomplishments of unit personnel, enhance individual productivity, and build unit morale.

Colonel Looney commanded an F-15 unit in Europe. He believed in always selecting his best pilots for the wingwide "top gun" award competition. After he took over, he reoriented the squadron selection process by instituting a unit fly-off (as opposed to a casual selection) to ensure his best pilots would enter the wing competition. This process required a little extra work, but the results—six out of six wins in wing fly-offs—were well worth the effort.[10]

The timing of wing awards should never surprise a squadron commander—there are few credible excuses for not submitting a unit member for an award at this level. Most commanders will tell you that their unit's nominee for a wing award program is always the result of a squadron program.

Squadron Awards

Squadron recognition programs give the unit commander a chance to be really creative. Some commanders institute unit awards programs to help solve a unit problem. Col Tom Whaylen instituted a "lab of the quarter" program in his technical training group at Keesler AFB, Mississippi, not only to recognize superior work, but also to upgrade the overall appearance and care of his group's facilities.[11]

Lt Col John Murphy presented many of his C-130 unit's awards, no matter how big or small, at squadron social functions. John believed in going public with unit recognition whenever possible.[12] Commanders also relate how they used the base newspaper extensively to broadcast unit awards. When you do this, always be sure to read the final proof for accuracy. Recognizing the achievements of unit personnel does not require a large production to be effective. Lt Col Bob Lawrence would personally meet every new C-141 aircraft commander immediately upon landing and completion of his or her first mission in command.[13] Many commanders relate that they built special recognition boards in their squadrons to display award winners. Colonel Gibeau wrote a short personal note to the family and parents of every award recipient from his unit. He received many thank you notes in return and quickly gained the respect of his people, who realized he really cared about their welfare.

Many squadrons use recognition programs to help focus the unit on mission accomplishment. Training squadron commanders tell me their units ran both "instructor of the quarter" and "airman/NCO/junior officer of the quarter" programs. Between the two programs however, the instructor program got the most emphasis from the commander—it dealt directly with the primary mission of the squadron. Other units automatically nominate their "instructor of the quarter" for the wing "airman/NCO/junior officer of the quarter" competition.

Regardless how large a squadron recognition program you develop, as commander you will need to keep a few basics in mind:

1. Keep it workable, fair, honest, and above-board.

2. Take an active interest in the selections and resulting awards.
3. Recognize your people in public as often as possible.
4. Use the program to help emphasize your priorities. If you can maintain an effective recognition program for all the categories discussed in this chapter, the odds of keeping unit morale high and mission accomplishment excellent are well in your favor.

The Bad--UCMJ and Discipline

Not every command function is a bed of roses. True story: Christmas eve/morning 1988, my wife and I had settled our daughters down for the night, finished placing the presents around the tree (assembled where necessary), consumed the milk and cookies, and finally collapsed into bed around midnight. We had six hours of sleep time available (max), five hours likely. I am awakened at 0215 by the ring of our friendly telephone. I pick up the receiver and the person on the other end says "Lt Col Timmons, this is Sgt Jones at the security police desk, sorry to call you at this time of the morning. We've just picked up one of your squadron members on base for possible DUI; if you or someone from your unit can get here soon, this one's on us, it's Christmas." My first thought is to order a quick execution, but my better senses quickly gain control and I reply, "Sarge, I'm inbound, give me about 15-20 minutes, thank you very much for your help." Two hours later I returned home, after retrieving an extremely frightened junior member of my unit from the security

> **PROVERBS FOR COMMAND**
>
> "Be people oriented."
>
> "You have to acknowledge people's intelligence."
>
> "Honesty—don't game your people."

police squadron and depositing him at his quarters where he would remain. The girls were up at 0530, ready to go—a very unusual Christmas morning!

Why did the security police call me and not the member's immediate supervisor? Why didn't I call the immediate supervisor or maybe my operations officer, who lived close to base? One simple reason—I, and nobody else in my squadron, had "*COMMAND*." Authority under the Uniform Code of Military Justice (UCMJ), is a very important, unique aspect of command and one that cannot be delegated—you hold the big hammer and you need to know how to use it properly.

Learning How to Use Your Authority

Most all former squadron commanders will admit they were inadequately prepared to handle discipline and UCMJ matters correctly when they first took command. All wished they had received some type of training in this area before learning the details by experience. Most Air Force officers have little contact with such matters before they're handed a unit flag, and the Air Force does not have a "pre-command" course. The types of discipline situations a commander can run up against range from tardiness through murder and *everything* in between—if you think you've seen or heard it all, you're dead wrong! More likely, after a year or so in command, you will firmly believe you grew up in a cocoon.

So, what can a commander do to become more knowledgeable on discipline and UCMJ matters before receiving the late night phone call or other "bad news" from unit supervisors? Most commanders recommend a proactive approach where the commander visits with the JAG, security police, CBPO, OSI, social actions, chaplain, family support, and the base hospital. The purpose of these meetings is to find out what role these units/agencies will play in various discipline matters, how these agencies relate to each other, how your unit and the agency can best coordinate actions and communications when needed, how they can help the commander and make his job easier, and how the commander can help them.

A couple other helpful TIPS before you have to dive into the discipline arena:

1. Find out what role the wing commander plays in this area; some take a very active role, to the point of dictating exactly what will occur in each specific case; others want squadron commanders to handle every situation totally by themselves; most are somewhere in between.

2. Investigate how the local authorities outside the main gate operate with regard to your people, military and civilian—their cooperation, jurisdiction, and relationship with the security police. This information is critical overseas!

Using Your Authority

Unless you are the luckiest squadron commander in the Air Force's history, at some point you will be presented a situation that will require your action in the discipline arena. The vast multitude of possibilities is much too great to cover in this book. However, in my conversations with commanders, a set of basic recommendations and principles to follow when faced with any discipline/UCMJ situation was mentioned time and again: Before acting yourself or directing action on a discipline matter, get inputs from everyone possible. Investigate thoroughly from all angles, and look for recommendations from the JAG and CBPO.

1. Use your chain of supervisors where and when applicable; they should be closer to the immediate situation than the commander. If the offender is enlisted, the squadron first sergeant is absolutely critical—closely listen to his or her thoughts and advice.

2. Keep those above you informed; don't ever let your immediate supervisor or the wing commander get blindsided by a discipline issue from your squadron; if you let this happen often, you will not have a squadron.

3. Document well by keeping a detailed record of interviews, counseling sessions, correspondence, and occurrences regarding any unit discipline action. By accomplishing this simple task, you will protect yourself, help yourself reach the desired conclusion, and possibly help those who follow in your footsteps.
4. Run it by the book, keep things fairly formal, take control of matters, and investigate all possible discipline and rehabilitative actions.
5. Confidentiality is important; these types of matters are not for public consumption.
6. After all the above—YOU make the final decision, you are the commander!

By following these six recommendations, you as a squadron commander can usually effect the desired outcome of the situation while at the same time protecting yourself and your unit.

Instead of attempting to discuss every type of problem a commander may face, here are just four examples (names and details slightly altered) of types of problem situations that can, and do occur:

Staff Sergeant Smith was a life-support technician at an overseas fighter base. He had previous problems controlling his weight and was an average performer. Prior commanders had let the ball drop regarding his weight problem, resulting in Sergeant Smith's never seriously attempting to correct the situation. A new commander inherited this problem and directed the life-support supervisor to counsel Smith, who was then put on the weight control program. Everything was run strictly according to the regulation—medical evaluation and all. After a month of unsatisfactory weight loss progress, the commander asked the first sergeant to talk to Smith about the situation and to emphasize the possible ramifications of a continued lack of progress. All counselings were documented.

A month later, no progress; the commander decided it was time to enter the fray. He called Sergeant Smith in for counseling and used a letter of reprimand. The commander emphasized that the squadron was trying to help the sergeant with the weight problem, but would not bend on the regulation or program itself. During the counseling,

Smith related that he was not reenlisting. Following this discussion, the commander contacted the CBPO and JAG to investigate possible actions. Smith's performance dropped below average, but his supervisor and the first sergeant could pinpoint no other outside problems.

Before any other actions occurred, someone in the squadron discovered that four of the seat kits in the jets had been sabotaged. The commander immediately called the OSI. Following an intense, but fairly quick investigation, Sergeant Smith was fingered. He was again called into the commander's office. This time the conversation was mostly one-way. After reading Smith his rights, the commander formally charged him under the UCMJ. Smith wisely asked to see the area defense counsel, and the session with the commander stopped there. The commander did everything by the book, honestly and above board. He involved the immediate supervisor and the first sergeant early, and he called the OSI as soon as the situation warranted. Smith ended up in prison.

Second Lieutenant Jones was an ROTC DG in pilot training. He was on top of the world—had a regular commission and was doing great through the T-37 phase of training, on his way to the F-15 he coveted. Shortly after beginning T-38 training, his lifestyle caught up with him. He was thrown from the back of a motorcycle at 0200 on a Saturday morning, on base, without a helmet. Both he and the driver were intoxicated. The driver, who had lost control, suffered three broken limbs. Jones was in a coma for four days—never to fly again, dreams shattered. He took two months to recuperate fully and was placed in a job at a wing agency awaiting a final decision on the line-of-duty investigation. His old squadron commander still had command authority over him and closely coordinated actions with the JAG. The line-of-duty determination was finally decided, nine months later in the Pentagon; the result—Jones had been injured "in the line of duty."

Lieutenant Jones was now clear for reassignment. His duty performance was excellent and none of his coworkers saw any problems or complications as a result of the accident. Jones received orders for a duty and location he desired.

Then, bang—three days before departing PCS, he got stopped on base for DUI. A new squadron commander, who had had frequent contact with the lieutenant, got the midnight phone call and couldn't believe his ears. Things got very formal at this point. The squadron commander served Jones with court-martial charges. Subsequent to dialogue between the legal office and the area defense counsel, Jones wound up with an Article 15 administered by the wing commander. Further, after initial interviews with social actions, Lieutenant Jones went to an alcohol rehabilitation center. This was the single best thing for him in years because it solved a problem!

After consultation with the CBPO, the commander delayed Jones's promotion to first lieutenant. This essentially gave him six months to prove himself worthy to stay on active duty. The commander backed the lieutenant in his efforts to prove himself, even though he had his doubts. The lieutenant worked for another wing agency but the same squadron commander still "owned" him. Miraculously, Jones pulled it off—his performance convinced both the wing commander and the squadron commander that his problems were behind. He was promoted, late, to first lieutenant. He soon was reassigned to another career field to basically start his career over. The commander played it fairly tough, but gave the officer the opportunity to pull out of a deep hole. He investigated all the possibilities and determined that a delay of promotion was the best action.

Airman Adams worked at individual equipment in the supply squadron. She was rapidly building a history of tardiness and poor performance. The unit had a strong first sergeant, who was very much attuned to his troops and kept the squadron commander informed on all matters. The commander let the first sergeant do his job.

Airman Adams and the Air Force were simply not cut out for each other. The first sergeant and Adams's immediate supervisor had done all they could to help remedy the situation—talks, counselings, advice, and so on. Finally, after a missed formation, the commander became directly involved and stayed involved through a series of actions that resulted in Adams's separation from the Air Force through 39-10 action.

The commander was very careful to protect his unit and himself throughout the entire process. He closely coordinated all actions with the CBPO and JAG and followed the script to the letter. He even ensured the presence of his secretary, a female, every time he formally counseled Airman Adams. Every detail was well documented. Beware—the type of disciplinary actions that do not result from criminal action can get fouled up quickly if the commander deviates from the book.

Major Moore was a rated officer working as a maintenance supervisor on the east coast. He was not actively flying, but did have a current private pilot's license. He was approaching retirement and was still bitter over his passover to lieutenant colonel. When he began applying to the airlines for a job after retirement, he decided to increase his chances of getting hired.

He forged government documents and lied to the Federal Aviation Administration (FAA) to obtain a commercial license plus a 737-type rating. The major was caught when one of the airlines called the base to gather more information on his background.

In this situation, the wing commander immediately placed the officer in a different duty on base and took control of the entire discipline/UCMJ problem himself. He rightfully recognized the gravity of the situation and saved the supply squadron commander a lot of work. After a long OSI investigation, a court-martial trial was finally scheduled. However, the squadron commander sensed something going wrong when he learned the major had gone to the unit orderly room to pick up a few old documents needed to outprocess from base and retire. The commander tipped off the CBPO, base commander, and wing commander who then ordered the officer to cease any retirement actions. The squadron commander had been relieved of any command actions in this case, but he was smart enough to keep abreast of the situation and help avert a potentially sticky situation. Major Moore "visited" Kansas for a while.

To conclude this section, I want to pass along a few short innovative techniques I learned concerning discipline and UCMJ matters.

LTC Steve Rippe recounts how threatening to write a young troop's father about a discipline problem will quickly get the individual's attention—they are deathly afraid of this action. Steve also told me how one of his solid troops had made a mistake and Steve decided not to "kill him" over the issue. So he signed a letter of reprimand, brought in the offender along with the offender's wife, tore up the letter in front of them, and gave the pieces to the wife for safekeeping. LTC Rippe had no more problems with that soldier![14]

More than one commander related how they used a "desk drawer" letter of admonishment/reprimand effectively. When they left command, they gave the letter torn in half in an envelope directly to the offender.

Command authority in discipline matters is a very serious matter. Overall, Colonel Lorenz had the best philosophy I've heard: "You are either going to shoot the guy or rehabilitate him; I haven't shot one yet."[15] I agree with his bottom line here—if at all possible, your job as commander should be to rehabilitate the individual first, to return the individual to being a productive Air Force member; if your best efforts there fail, run things by the book and continue to look out for the best interests of the Air Force.

PROVERBS FOR COMMAND

"Kiss the good guys and kill the bad guys."

"Be consistent and set the example."

"Treat people evenly across the board; if you don't they'll quickly pick up any inconsistencies, especially your enlisted personnel."

"Be prepared to wear black hats and white hats."

"Get all sides of a story or situation before acting."

The Ugly—Times of Crisis

Midnight phone calls to a squadron commander are rarely good; most bring the type of bad news discussed in the last chapter—discipline or UCMJ matters. As unpleasant as these calls may seem, however, they pale by comparison with a phone call that informs the commander of a squadron member's death.

This section will discuss what I call "times of crisis"; situations of a tragic and very unpleasant nature that usually touch the emotions or lives of most squadron members. Some examples of this type of situation are aircraft accidents, other accidents resulting in death or serious injury, the death of a member's relative, and closing down the squadron or a massive change in unit mission and personnel. A commander's intense involvement in these situations can last for months or just a few days. Regardless of how long or short the involvement, the commander will remember the experience the rest of his life—this also "comes with the turf."

Many commanders never suffer a significant crisis during their tenure. While the possibility of a member's death or an aircraft accident is always in the back of a commander's mind, he or she is never really ready when it occurs. The following are examples and tips on how commanders handled crisis situations. Their methods are certainly not the only ways to handle similar situations, although they all have one thing in common—in each situation, the commander met the situation head on and personally worked the crisis.

Aircraft Accidents

In early April 1975, President Ford announced Operation Babylift, an effort to airlift babies, children, and relatives out of South Vietnam, whose downfall was about a month away. The first mission out of Saigon in Operation Babylift was a C-5A from Travis AFB, crewed by the 22d Military Airlift Squadron (MAS). The plane departed Saigon for Clark AB with an augmented crew plus additional loadmasters. More than 70 passengers occupied the rear troop compartment and approximately 200 passengers (mostly women and children) were downstairs in the cargo compartment. On climb-out just past 20,000 feet, the C-5A experienced a failure of the rear pressure doors. This caused a massive depressurization and damage to multiple cables, leaving minimal use of any of the plane's control surfaces.

Miraculously, the pilots managed to turn the plane around using differential power and head back toward Saigon. Tragically, the plane hit short of the airfield, bounced over a river, split into three pieces, and crashed into a rice field. That day Col Charles Geer, commander of the 22d MAS received "the phone call" informing him of the accident. By evening the accident was the lead story of every television, radio, and newspaper in this country and most of the world. Colonel Geer was about to go through weeks of emotional hell that would test the leadership of anyone ever called commander. As a first lieutenant in the squadron, I witnessed his actions firsthand—I have never seen anyone handle such a situation better. For this and many other reasons, he remains the benchmark against which I judge all other squadron commanders, including myself.

Details of the accident were sketchy at first, so Colonel Geer did not attempt to disseminate information of which he was not certain. What he did know was that six of his crewmembers were missing and a couple were injured, one critically. After initially notifying the families of the crewmembers, he gathered all the information available, then began squelching rumors and basically organizing the squadron "family" to cope with the tragedy.

Foremost on his mind were the families of the one critically injured and the six missing crewmembers. The injured member had been flown to a hospital in Honolulu where he eventually died. Colonel Geer personally arranged for the family to fly to Hawaii where they arrived before the member passed away. The squadron passed the hat to pay for the tickets. It took six days after the crash to determine that the six missing members had, in fact, died in the crash. During this week, Colonel Geer personally visited the families of all six, daily and some times more often. He handled each visit professionally—with compassion, honesty, and sincerity.

During this entire ordeal, Colonel Geer never forgot the rest of his squadron. He was always straightforward in answering the many questions surrounding the accident and he consistently kept control of the situation.[16]

Lt Col Dave Knowles, commander of the 644th Bombardment Squadron at K. I. Sawyer AFB, also received a phone call about an aircraft accident. The call came at night and informed him that one of his B-52s, with eight crewmembers on board, had crashed on base. He first went to the command post and gathered all available information, the most crucial being that all eight of his crewmembers had been rescued from the plane by the excellent work of base rescue personnel. The injured were being transported to the base hospital.

Colonel Knowles quickly organized his "unit support network." He gathered his key squadron personnel, their wives, and each of the eight crewmembers' wives at the squadron. He arranged for other unit personnel to perform babysitting duty. Once all were present he came straight out and told everyone exactly what he knew and handled the various emotional reactions to the news. He also placed personal calls to the parents of each crewmember to provide straightforward information before CNN got the news. After the crewmembers were stabilized, he arranged for their wives to visit them at the hospital. In the days that immediately followed, he continued to concentrate on the families' needs via the squadron network.

Colonel Knowles was also concerned for his unit as a whole. He relates how the commander should watch the rest of the unit members carefully and not let such an accident affect their ability to perform the squadron's mission. To handle this aspect of the crisis, Dave immediately scheduled a commander's call and asked many base agencies to attend, including chaplain, flight surgeon, and mental health. There, he told the entire squadron all the details, had the agencies discuss a wide variety of related subjects, and began steering the unit back towards the mission.

Right after the commander's call, he held an informal gathering where everyone could "let loose and talk about it." He stayed with many of his squadron personnel for hours until they had "gotten it out of their system." This proved invaluable and allowed the unit to rapidly refocus on the mission.

However, this did not mark the end of Colonel Knowles's involvement. As he soon found out, the Air Force is well geared to support those who can do the mission and has effective procedures to deal with active duty deaths. The Air Force is not nearly as well prepared to take care of those who suffer major injuries and require transition to temporary disability status, when the Veteran's Administration (VA) takes over medical care. Few commanders know how the VA works, and the transition from active duty is traumatic for member and family. Gaining this knowledge requires time and research. A commander may easily become frustrated while he climbs this learning curve.

In retrospect, Colonel Knowles handled this crisis head-on and kept control at all times. He used a two-prong approach where he first took care of the immediate needs of the involved crewmembers and their families and then took care of the needs of the squadron as a whole, ensuring that they stayed focused on their mission.[17]

Lt Col John Wilcox experienced a similar situation as commander of a B-1B squadron at Dyess AFB, Texas. His unit lost crewmembers in an aircraft accident that also generated a testy investigation board. John was very careful in how he, as commander, approached the unit and the board. He recounts he had to back off just a little bit emotionally, not an easy task. He also warns that a commander can't get too personally involved with an accident investigation board. He recommends meeting such a situation head-on, controlling the action where you can, but being slightly detached emotionally to best serve the unit and the Air Force.

Deaths and Serious Injuries

The toughest task for a squadron commander to perform is informing a spouse and family of the death of a unit member. A couple commanders I interviewed accomplished this unpleasant duty and offered some tips:

1. By all means, this is your duty; others may accompany you but you need to be there.
2. Be prepared to answer a lot of questions about what happened and what will happen in the future—have the answers ahead of time and/or the necessary contacts ready to swing into action.

3. When you first speak, be straightforward and get it out quickly; when the family sees you at the door in a Class A uniform along with the chaplain and others, they will know something is wrong; you are likely to get an emotional response even before you start talking.

4. If possible and applicable, ask a close friend of the family from the squadron (member and/or spouse) to either accompany you or follow you by a few minutes.

5. Don't be surprised at the initial reactions of the family—people's emotions will often fool you; don't let this prevent you from handling the situation in a professional and compassionate manner.

After informing the family of a squadron member's death, the commander then needs to quickly appoint a few people to carry out some important duties and then oversee, not micromanage, the necessary tasks. A few more TIPS here:

1. Appoint a sharp, well-organized summary court officer, one who knows the family and who the family will trust.

2. This summary court officer will need to visit many base agencies seeking help. Grease the skids by calling the agency commanders or chiefs to let them know who's coming in their direction.

3. Appoint people other than the summary court officer to handle any additional necessary arrangements, such as memorial services, funerals, transportation of the deceased and family, and so on.

4. Keep your boss informed as things progress.

5. Periodically stay in touch with the family, ensuring that all their questions are answered and all their rights and benefits have been fully explained.

6. If possible, personally inform the member's close coworkers; try not to allow them to hear the bad news through the rumor mill.

Certainly each individual situation will present a commander with different challenges, but I believe a commander who follows the above suggestions will guide the actions as smoothly as possible under the circumstances.

Informing a squadron member that a close relative has passed away is another unpleasant commander's task. This task is often closely followed by working arrangements to allow the member to conduct personal business.

Early one November afternoon in 1988 I received a call from the Red Cross informing me that the father of one of my second lieutenants had died of a sudden heart attack that morning in Saint Louis. I told the Red Cross that I would handle the notification and would call them back after I had talked to the lieutenant. I quickly cross-checked all available personnel data, ensured that the Red Cross had the correct individual, and then asked my operations officer to go get the lieutenant (a student) out of a simulator mission and escort him to my office. Meanwhile, I called the chaplain's office and asked the people there to stand by if I needed assistance.

Once the lieutenant arrived and we were in the privacy of my office I got straight to the point—I told him of the call from the Red Cross and that I was sorry to inform him that his father had died earlier in the day. The lieutenant was shocked—this was totally unexpected. I left him alone in my office to collect his thoughts and then set my executive officer and the lieutenant's flight commander into action checking flight, loan, and schedule possibilities.

After the lieutenant calmed down, I suggested he talk to his mother back home. This AUTOVON call I initiated myself. I called the operator at Scott AFB, Illinois, explained who I was, what the situation was and asked her to connect me with the off-base number. Once I heard the phone ringing, I gave it to the lieutenant and let him use my office in private.

Upon completion of the call, I told the lieutenant not to worry about his training program and to take as much time as he needed back in Saint Louis. His leave form was ready to go, plane reservations confirmed, and a loan, if needed, already approved. We also had ground transportation lined up. I received the call from the Red Cross at 1345 and by 1630 the lieutenant was airborne for Saint Louis. The squadron had really come together quickly to help a member.

Closing a Squadron

Colonel Gibeau was commander of the 71st Tactical Missile Squadron (GLCM) at Florennes AB, Belgium, when the signing of the Intermediate Nuclear Forces (INF) Treaty doomed his unit to closure. This type of unit crisis, although not as tragic as a personnel death, will still be a very emotionally difficult event for a commander—the squadron is his baby, and someone or something is taking that away!

In Colonel Gibeau's case, closing down the squadron meant turning in every piece of property to USAFE, actually hauling the building away and reassigning all unit personnel. This was accomplished for every unit at Florennes, and the base was eventually closed.

> **PROVERBS FOR COMMAND**
>
> "In times of crisis you will feel despair and self-analysis; the unit needs your leadership at that moment more than ever."
>
> "Integrity."

John's two priorities for unit closure were working assignments for all his personnel and accurately accounting for and turning in all unit equipment.

Colonel Gibeau relates that he took a personal interest in each and every personnel reassignment action. This required a great deal of time and effort coordinating actions with USAFE and SAC headquarters. He felt fairly successful in this task because he took a proactive role and knew the desires of his people. In working the equipment turn-in, Colonel Gibeau followed the guidance of his wing commander by organizing the actions through his unit flights and keeping a close handle on all necessary actions and paperwork.

It was a sad day for John Gibeau when they carried off the squadron building. He did have the satisfaction of knowing the closure had been accomplished professionally and maybe he and his people were "simply irreplaceable."[18]

A crisis rapidly forces a commander to center stage. If faced with such a situation, a commander needs to set courses of action early and remember to keep the unit informed and focused in the right direction.

The Role of Spouses and Families

The role of spouses and families in an Air Force unit is an emotionally charged issue that received a great deal of high-level attention during the 1980s. On 3 August 1987, the *Air Force Times* reported a story of two Air Force wives who were pressured to quit their jobs in order to participate in base activities. As a result of this highly publicized incident at Grissom AFB, Indiana, the secretary of the Air Force formed a blue-ribbon panel to study the role of the military wife. The issue at stake in the incident and study was the historic pressure on the Air Force wife to support the Air Force family, even if it meant giving up other things in her life or doing things she really didn't enjoy doing. The resulting report and policy statement, which were published in 1988, emphasized the right of a military spouse to pursue her own ambitions and goals without any pressure from the Air Force. It also stated that the Air Force welcomed any volunteer support given to the various support activities around our bases.

Air Force leadership met this question head on and came up with a good policy that reflected the society in which we live today. Yes, the spouse and family do have an important part in Air Force life. How this policy gets translated in the field is up to you, the squadron commander, and the commanders above you. During my interviews, I asked the commanders about the role of spouses and families in their squadrons. I received a wide variety of answers on this issue, and it became quite evident the Air Force policy is being implemented (or not implemented) in many different ways. One commander said, "The Air Force is fooling itself if they think verbiage from the secretary's office is changing the way wives are used in the wing." This officer still saw a lot of pressure on wives coming from above him.

This section will not attempt to give any hard and fast answers to questions about the role of spouses and families in an Air Force squadron—there are no set answers. Each Air Force member, each squadron, each wing, and each mission creates a different set of needs. Isolated stateside locations, overseas tours, large metropolitan areas, alerts, extended TDYs, unit deployments, and classified missions all present military commanders different sets of circumstances and problems with which to deal. In most cases, the squadron commander is at the center of each situation and is the key player in helping to solve problems and meeting needs so that ultimately the mission can be accomplished. As a commander, you should remember that you will be dealing with different generations of people who have different values. This is true not only for those in your unit but also for those above you in the chain of command! We'll start real close to home for most commanders.

The Squadron Commander's Wife and Squadron Spouse Groups

Your taking command of an Air Force squadron will affect your wife and your family, if for no other reason than the added requirements the job places on your time and your activities. Your wife will no doubt ask you many questions concerning what your new duty means for her—"What is expected of me?" The answer will vary with each unit and base. RECOMMENDATION: Find out *early* exactly what is expected of her, get the lay of the land, and understand how the wing and base operate with regard to spouse matters. The primary object is to ensure that your wife does not get blindsided by something she is not prepared to handle.

Is pressure still out there? In some locations yes, other locations, no. One commander told me how his boss's wife, still of the old school, brought subtle pressure to participate on the squadron commanders' wives, and generally created a tense atmosphere. Another commander related how he was told he had been selected to command a squadron. He was in a meeting with the wing commander, vice commander, and deputy commander for operations when the wing commander said, "Congratulations, you've been chosen to command one of the flying squadrons, and oh, by the way, your wife will be quitting her job, won't she?" These types of situations are certainly in the minority these days, but they still exist.

Once you obtain the answers to your questions, you may want to discuss the situation with your wife and figure out what type of involvement in the unit she would like to pursue. Some commanders' wives enjoy the traditional role of leading the unit spouses in all kinds of activities, while other commanders' wives are uncomfortable in this role. One commander related a very good piece of advice—a commander's wife should do whatever she is comfortable doing and, where possible, the priority should always go to the squadron over other agencies or groups, such as the base OWC.

No matter how your wife approaches the situation of being a commander's wife, she will need some help—mainly from you! It's not a bad idea to explain to your people how your wife intends to function. At one of my first staff meetings, I told everyone that my wife was interested in pursuing activities the group was interested in doing, on a schedule the wives' group wanted. She was not interested in a meeting for the sake of having a meeting. I also said, "Mrs Timmons" is my mother, and my wife is not a lieutenant colonel, she's a "Linda."

With regard to the commander's and commander's wife's role in unit spouse groups, the following are some recommendations and pieces of advice from other commanders:

1. Let your wife run things her way and generally stay out of her way! However, be available to help when needed; your position as commander can quickly facilitate certain actions.

2. Don't force or push for participation; any participation obtained through pressure always turns out to be poor participation.

3. Don't link a wife's activity to a member's career progression—don't count noses!

4. Find out what people like, what's pertinent, and adjust spouse group activities to the needs of the people. Offer choices; one new squadron commander's wife sent a written survey to all of the unit spouses to get this information, responded to the results, and ended up with terrific participation.

5. Many commanders periodically attend spouse meetings to explain the unit's mission and mainly answer questions. You have to be careful here. One strategic airlift squadron commander was asked why the crews had to pay out all this money for food and lodging while on trips; when he replied "That's what TDY pay is for," the wife then asked, "What's this TDY pay? I've never heard of my husband getting any money for these trips!"

6. As commander, don't forget to thank the spouses for the work they do. Colonel Ron Love made it a point to stop by the spouse meetings whenever he could just to say thanks and let them know someone really appreciated the volunteer work they performed.[19]

7. Beware of potentially sticky situations. I heard of one outgoing squadron commander's wife who didn't want to "let go," which made it difficult for the new commander's wife to get things running in her direction.

8. Understand how your unit's demographics can affect spouse groups:

a. In some locations, enlisted and young officers' wives have to work just to pay the bills; they will have little time for anything else regardless of their desire.

b. If you have a lot of young couples in the unit, they most likely will be starting families—young children require a great deal of time, remember!

c. Overseas locations tend to have closer-knit groups; bases near large metropolitan areas tend not to have such close-knit groups; some bases have a large number of "homesteaders" who have close ties to the local area.

d. There is often a need for spouses to "pull together" in situations where the squadron deploys for extended periods of time or where personnel are often gone on many short duration TDYs; during this time the commander's wife will be the focal point of most activity and telephone calls.

9. I heard of one commander and his wife who had planned in advance what actions they would need to take in the event of a death in the unit; he had his plan for handling the squadron members, and she had her plan on how to work with the spouses to help those in need. Sure enough, the game plans had to be used. The commander told me this prior planning saved a great amount of time and needless suffering; he felt he would have been lost had he not thought through the potential situation before it occurred.

10. Maj Bill Moseley told me that representatives from his spouse organization are part of every decision group that discusses any unit social or moneymaking project. He says this action brings the spouses closer to the unit and results in successful projects.

11. BE YOURSELVES!

Involving Families in the Squadron

Many commanders believe that the more they can involve a member's family in the squadron, the better the member's duty performance. They point out that involving the family in the squadron does not necessarily mean loads of activities, but rather means ensuring the spouses and families understand what the squadron is all about.

One commander tells of holding a "career day" where families could come to the squadron and talk to all available members, from the youngest airman to the commander, about that individual's particular portion of the mission. This enabled the families to better understand how the squadron was put together and how it functioned. The day also gave everyone a better appreciation for the deployments and exercises in which the squadron periodically participated.

Another commander told me that he or his first sergeant would always call the spouse of any member who was off on a long TDY to ensure things were going well, and if there were any problems, he made certain things were fixed or corrected before the member returned. He placed a premium on taking care of his entire squadron family.

Lt Col Kathy LaSauce commands a large aerial port squadron at Andrews AFB, Maryland. She invites spouses and families to all of her commander's calls. She understands the demographics of her unit and how hard it is for her people to all make a meeting at the same time—so, she makes it easier for them. She feels this builds better unit cohesion, allows the families to become more knowledgeable about the squadron, and gives her the opportunity to address a wide variety of issues that everyone needs to hear.[20]

On a lighter side, involving families can start early. One commander and his wife would always give presents to new babies in the squadron. He thoroughly enjoyed going to the hospital to visit the newest unit members. Being the commander also allowed him to "stretch out" normal visiting hours.

What can happen when families are not involved at all in a squadron, when families are unaware of what the member does, or have little knowledge of what the mission is all about? One extreme case was related to me by a commander whose unit members infrequently went TDY, but when they did, it was usually for a period of 90-120 days. This commander had a young married NCO who had been on station more than five years and had also grown up in the local area. The NCO was selected for a certain 120-day TDY because he possessed the required skills and was next in line to fill the unit's commitment.

The very next day after the TDY orders were published, the commander received a call, at home, from the NCO's wife wanting to know why her husband was being ordered away for four months and stating that the NCO couldn't go for a wide variety of reasons! The commander asked the wife and the NCO to come in the next day and he would explain. The conversation during this meeting revealed the NCO's wife to be basically unaware of what her husband actually did in the Air Force, how the squadron functioned, or what comprised the wing's overall mission. Commissaries, base exchanges, free medical care and all were great, but what was this thing called "mission"? This commander believes that all too often Air Force families, and sometimes the members themselves, forget what military service is all about. To prevent future similar occurrences, the commander embarked on a family involvement program aimed at educating the squadron families on the Air Force and the squadron.

The Ideal

What qualities make up an "ideal" Air Force spouse/family? I'll end this chapter by giving my opinion on the ideal characteristics for a spouse/family.

1. The ideal spouse/family understands and appreciates the member's duties and the unit's mission. This does not mean they have to possess any kind of detailed knowledge, just a basic awareness of matters.
2. The ideal spouse/family is supportive of the member. They do not necessarily have to be cheerleaders 100 percent of the time, applauding every accomplishment; too much of this will go to the member's head!
3. The ideal spouse/family accepts the particular hardships that come with military service. They don't have to like the frequent moves, TDYs, remotes, and alerts, but they should understand why they occur and accept the possibilities.
4. The ideal spouse/family participates in activities and organizations they enjoy, when they can afford the time. If a wife wants to run the wives' club because she enjoys the activities, great; if she does it solely to enhance her husband's career, not great at all!
5. Finally, the ideal spouse/family functions daily with the efficiency that allows the member to concentrate fully on his or her duties and ensure mission accomplishment. The best thing a family can do for a military member of any rank is to let him or her do the job!

If, as a squadron commander, you had an entire unit full of families of this type, your problems would be few and far between. Realistically, you will not have a squadron composed solely of these people. Rather, you will have a wide diversity of spouses and families who will most likely present you with some challenges during your tenure at the helm. Hopefully this chapter has provided some thoughts and ideas to help you better involve spouses and families in your squadron.

Notes

1. Lt Col Craig Koontz, interview with author, Ft McNair, D.C., 29 November 1989.

2. Lt Col Bill Looney, interview with author, Ft McNair, D.C., 24 October 1989.

3. Lt Col Kathy LaSauce, interview with author, Ft McNair, D.C., 14 March 1990.

4. Col Bill Welser, interview with author, Pentagon, VA, 27 October 1989.

5. Lt Col Steve Lorenz, interview with author, Ft McNair, D.C., 16 October 1989.

6. Lt Col Dave Morrow, interview with author, Ft McNair, D.C., 21 September 1989.

7. Lorenz interview.

8. Lt Col Ron Love, interview with author, Ft McNair, D.C., 1 November 1989.

9. Lt Col John Gibeau, interview with author, Ft McNair, D.C., 28 November 1989.

10. Lt Col Bill Looney, interview with author, Ft McNair, D.C., 24 October 1989.

11. Col Tom Whaylen, interview with author, Ft McNair, D.C., 3 October 1989.

12. Lt Col John Murphy, interview with author, Ft McNair, D.C., 19 October 1989.

13. Lt Col Bob Lawrence, interview with author, Ft McNair, D.C., 2 November 1989.

14. LTC Steve Rippe, interview with author, Ft McNair, D.C., 2 November 1989.

15. Lorenz interview.

16. Col Charles Geer, personal observation, Saigon, South Vietnam, April 1975.

17. Lt Col Dave Knowles, interview with author, Ft McNair, D.C., 15 November 1989.

18. Gibeau interview.

19. Love interview.

20. LaSauce interview.

Chapter 4

Communicative Leadership

As stressed throughout this book, a squadron commander's focus should be mainly on his unit's mission and people. Having already covered these topics in earlier chapters, I now want to concentrate on some ins and outs of a commander's daily activity by discussing the "glue" that holds the squadron together—communication. Good communication is the key to effective squadron leadership—of any squadron, at any location!

PROVERB FOR COMMAND

"You must command. To do so, you must control. To control, you must have communications. Because of change, you're dead without intelligence." —Gen Colin Powell, then a corps commander, talking to his unit leaders.

Communication is not always an easy task—ensuring that everyone in the squadron gets the complete, unfiltered word is one of the toughest tasks for a commander. Further, consistently receiving important feedback and advice back up the chain to the unit commander is also a "tough nut to crack."

Communication takes many forms and can be divided into multiple subcategories—written, verbal, internal to the unit, and external to the unit. A commander will likely find verbal communication to work best in some situations and written communication in other situations. Verbal communication tends to be faster than written communication, while written communication is generally more detailed and permanent.

Communication will "happen" in any squadron; however, if the boss is not careful, it may not happen the way he desires. No commander wants the "rumor mill," where incorrect information is often passed along by unknowing individuals as fact, to be the primary means of communication.

Meetings

Among the numerous methods commanders use to get the word out to their people, meetings are the most common.

Many commanders whose squadrons had *daily meetings* for shifts, crews, or sections often used these meetings to talk personally to large portions of their unit at one time. Each flight commander in Lt Col Ray Chapman's pilot training squadron started out the work day with a mass flight briefing. If Ray had something really important he needed to discuss with all his instructors and students, he would personally attend each flight's briefing.[1]

> **PROVERB FOR COMMAND**
>
> "Ungarbled communication is an important key to effective operations."

Lt Col Steve Lorenz's KC-135 crews would flight plan every morning. The schedule was designed to have each crew plan every third day. By attending these planning sessions for three days in a row, Steve was able to reach most of his unit in a short period of time.[2]

All of the branches in Lt Col John Bell's maintenance squadron would form up each morning before beginning their duty. These formations provided John and his senior supervisors the opportunity to talk to the entire unit during the course of the day. Two maintenance squadron commanders at McGuire AFB explained how they used this technique to get the word out to all their people even though their squadron worked in three shifts, 24 hours a day.[3]

The *weekly staff meeting* is probably the most commonly used method to disperse information in a squadron. It provides the commander the chance to discuss important issues with his squadron staff and to provide guidance and direction. However, in order for the information to reach the lowest level in the unit, the staff and supervisors must accurately pass on the commander's words to the troops.

One of the problems I experienced with my weekly staff meeting was not having my key supervisors, especially my flight commanders, attend on a regular basis. Our training schedule was somewhat erratic from week to week and holding a staff meeting at the same time each week caused significant problems for my schedulers. I strongly believed in the mission's having top priority, but I also needed my key personnel at this meeting. So, I let our training mission schedule drive the staff meeting time and scheduled the staff meeting when we could get the best attendance. Effective communications sometimes requires flexibility.

Weekly meetings need not be limited to your staff. Some former commanders of smaller squadrons told me they held weekly meetings for everybody. This is a good technique if your unit's size and schedule allows you to gather everyone at once.

Lt Col Ron Love used his *quarterly planning meetings* to help facilitate communication in his C-9 squadron, where a large number of TDYs made it difficult to get his key staff together on a regular basis.[4] Lt Col Bob Lawrence held periodic *certification boards* for newly upgraded C-141 crewmembers and often used these forums to get the word out to a small portion of his unit, who would in turn pass along the boss's word to the rest of the unit.[5]

Commanders whose units deploy to other locations, such as Lt Col Dave Morrow's F-15 unit, often gather everyone together a few weeks before the deployment for briefings. They use these *special meetings* to discuss a wide variety of topics in addition to the upcoming deployment.[6]

Outside the Squadron

Squadron commanders spend a good deal (most commanders will say too much) of their time each week involved with meetings outside their own unit. These meetings come in all sizes, shapes, and forms and their importance to the squadron can vary greatly. Most of the meetings are mandatory—a specific individual or at least someone has to attend, as dictated by some regulation, policy, or higher authority. A lot of these meetings are periodic in nature—and hence provide a great deal of advance notice. Other meetings may be called on short notice. Regardless, as the squadron commander, you are responsible for ensuring your unit is represented at all of these gatherings. You may pass the authority through your operations officer, supervisor, deputy, or superintendent to your schedulers, but you can't pass the responsibility.

Time for a system to ensure your unit stays clean! Setting up these systems can become old and cumbersome, but if one works for you, stick with it. The best one I heard about here involved internal squadron cross-checks. Whenever the commander received a meeting notice, he immediately gave it to his operations officer who supervised the unit schedulers. The operations officer then made certain the individual who needed to attend got a copy of the notice, along with the schedulers. The ops officer also kept a large calendar in his office where he recorded all upcoming meetings. If the commander needed to go to the meeting, the squadron secretary noted it on the commander's calendar.

The commander and operations officer would spend two minutes each Friday reviewing the large calendar for the following week and then cross-checking it with the unit's scheduling boards. Easy to do and pretty fail-safe. This commander related he had few if any problems with meeting attendance after establishing this system.

That takes care of getting to the meeting; what do you do when you get there? Simple, you participate, right? Yes, but it helps to be prepared, especially if the meeting is being run by your boss or someone else up your chain of command! To be prepared, a smart commander will ensure he knows the agenda before the meeting.

This doesn't mean just knowing the topics to be covered but also knowing why the meeting is being held and what is really on your boss's mind. Use whatever means you have available to gather this information and these meetings will be easy. Remember the commander discussed earlier whose squadron personnel had moved on to become wing-level execs. I bet he didn't have any trouble getting this information! Another thing to remember about meetings called by other agencies or units—it may very well be the most important occurrence of their day or even their week. Don't be surprised if those running the meeting believe it to be far more important than you believe it to be. Hopefully, they will understand this fact and will conduct a concise meeting that covers the necessary points but does not needlessly waste people's time. In this same vein, many agencies will want commanders to attend. As the commander, you know your schedule and your priorities best, so you should decide who attends, based on existing policies/regulations and what is best for the squadron. Sometimes people outside your unit need to be reminded of your priorities. Personal anecdote:

About ten months after I took command of my training squadron, my weekly calendar began to get crowded by all sorts of meetings called by wing agencies. Each agency, supported by my boss, wanted the unit commanders in attendance at their specific meeting. The problem was that nobody, except other commanders in the same situation, saw the overall effect on my schedule. I was left with little time to perform what I believed were my more important duties—those directly connected with our mission, such as administering flight and simulator progress checkrides to students experiencing training problems. Meetings were crowding out the MISSION!

Instead of just complaining to my boss, I first came up with a solution and presented it at his staff meeting. My boss's staff meeting was always held on Wednesday, so I suggested that any agency desiring to hold a meeting with the squadron commanders do so on Wednesdays, and leave the remainder of the week "meetingless" so the commanders could perform their primary mission. No one had any problem with this set-up and it worked well thereafter.

Commander's Call

Another vehicle for facilitating communication is the commander's call. This is a more formal meeting for the entire squadron, and its timing usually depends on the commander's desire. Some commanders hold commander's call each month, some every other month, and some less often. The average is about every other month.

Commander's calls are used for a variety of purposes, such as awards, decorations, recognition programs, guest speakers, films, and briefings. They allow the commander to put out a lot of information, ideas, thoughts, views, guidance, and direction to most all unit personnel at once. Many commanders believe their periodic commander's calls to be their most important meeting.

For this reason, experienced commanders will emphasize the importance of thorough preparation for these affairs. A couple of commanders related how they would completely dry run their commander's calls the day before; many others told me that they would at least have their exec or secretary make up detailed note cards for them to review before the event.

Commander's calls can be full of surprises that will test a leader's calm under fire. I remember once seeing a squadron commander get totally blindsided by the unannounced appearance of the wing commander at one of these calls. Fortunately the squadron commander/lieutenant colonel was slick of tongue—he immediately added the colonel into his current comments and smoothly introduced him to the squadron before the colonel had even made his way to the front of the auditorium. I was surprised once myself by a "birthday greeting" entering one of my commander's calls—I tried to act calm and cool while everyone got a good laugh.

Colonel Bell used an interesting technique with his commander's calls to answer people's questions. Realizing that many young troops are reluctant to stand up and ask the commander a question in a large gathering, John invited everyone to write down any question on the back of his or her attendance slip. He then ensured that each question got answered, either individually, through a staff meeting where the word then got passed on down to everyone, or at the next commander's call.[7]

One-on-One

Some of the most effective communication takes place during individual counselings or meetings. Many commanders place a premium on their first welcome meeting with a new squadron member and often try to schedule the meeting so that the new member's spouse can also attend. Col John Brooks, who commanded a C-130 squadron at Yokota AB, Japan, also tried to have his wife, Gail, at the meeting.[8] This type of welcome can really put the new member and his family at ease in a strange surrounding. Some commanders say they use this welcome meeting to impart squadron and/or personal philosophy to the new member. This helps everyone get off on the right foot and usually answers many of the questions that result from a PCS move.

Remember that the physical setting for any one-on-one meeting or counseling session involving the squadron commander can influence the effectiveness and outcome of the session. As the commander, you should have total control here. If you want the meeting to be very informal, you may want to go to the individual's workplace or meet somewhere outside the squadron.

If the meeting is to be held in your office, where you and the other person sit can influence the degree of formality of the meeting. One former PACAF unit commander had a set of personal guidelines he used for one-on-one meetings in his office:

1. For an informal meeting, the commander and the individual would sit at a table or on two similar chairs.
2. For a slightly more formal meeting, the commander would sit at his desk and the other person would sit somewhere facing the desk.
3. If the meeting called for a strict formal atmosphere, the other person's chair (optional) was directly in front of the commander's desk.

This same commander also gave me another TIP on how to set the tone for a meeting: most of the time the visitor was asked to "come in" or "go in" the commander's office, but on certain occasions, he or she was told to "report in" to the office. These simple phrases, usually spoken by a secretary or executive officer, can help the commander set the tone for the meeting.

A commander can also facilitate squadron communication by using a "special agent." Often this is the commander's executive officer or secretary whom squadron members often quiz concerning the boss's current thinking. I can remember that many times if I wanted some information to get out quickly, I'd tell my exec to put out the word and sure enough, by the next day, everyone had the word.

As good as the above-mentioned communication techniques may be, they all take a second row seat to the commander's actually getting out from behind the desk, walking around the squadron and *talking to his or her people*. I think LTC Steve Rippe, said it best: "The art of command involves knowing where and when to go down and get involved in a lot of details." Steve was big on getting out and talking to his people.[9]

Proverbs for Command

"Don't succumb to the Ivory Tower, get out in your squadron."

"Always have enough time for your people."

A tanker unit commander was right on the mark when he said, "It's much better to talk too much to your people than too little. They need to get to know you better." Your people do need to get to know you better, and I think that most former commanders will admit that they probably should have worked harder at this during their tour.

As mentioned before, *where* you talk to your people can be important. Taking the time to go to the troop's work station can provide a big boost for morale. *When* is also important—the middle of the night, on the flightline in 100-degree temperature, or at the front gate of the base in two feet of snow—not always when it's just convenient for the commander.

Feedback

Communication is a two-way street; a commander cannot operate in an information vacuum. All of the methods discussed above for giving out information can work in reverse. Any type of meeting worth its salt will involve communication to and from the leader. However, if you really want to find out how your troops are doing, seek them out on their turf and spend some time *listening*. The other, larger forums can be effective for getting feedback, but going directly to the individual is the best. I often used time after a Friday night flight or simulator mission to kick back, have a beer, and just talk with my instructors and students.

A true open door policy also works well. Once your squadron personnel realize you are willing to stop work and take time to listen to them, they will begin to frequent your office. You have to be careful here—once you announce an open door policy, you have to follow through and not put off anybody who may stop by with a problem or information. This will likely make it more difficult for you to finish paperwork or projects, but your people deserve your time. One commander related how he worked it: if his door was physically open, it was "open"; if the door was physically closed, he actually needed the time and privacy. He closed the door only when absolutely necessary.

Sometimes, no matter how hard you try, people are just reluctant to talk to the boss about certain matters, but are perfectly willing to talk at length to their own peers. Here again, a special agent comes in handy—someone who has the ear of everyone, the informal leader, the old head whom everyone turns to for small problems. Many commanders related how they often asked this type of person to help keep them informed on the unit's "health and morale." Commanders shouldn't operate in a secret or clandestine manner, but more in an open way, acknowledging the fact that some things just don't naturally float up to the boss.

Prioritize Routine Paperwork to Go outside the Unit

There are many "pulls" on a squadron commander's time during the day. Most commanders will rate attending to the mass of paperwork piled up in their in-box as one of their least favorite activities. Commanders would rather spend most of their time with their mission and people. To help ensure that you, as commander, have more time for the duties you prefer to do, I offer three suggestions:

- Prioritize your work.
- Develop efficient "systems" to effectively work those things you may not particularly like to do.
- Be proactive with your personal schedule.

Through the commander's in-box travel most pieces of paperwork, either coming into the unit from external units/agencies or going out of the unit to these same locations. The exact volume of paperwork the commander sees should be up to the commander! Most units have some type of administrative personnel who screen correspondence going to and coming from the commander's office—a secretary, executive officer, or administrative clerk.

As a new commander, you may want to see 99 percent of the paperwork. However, as you become more comfortable with the environment, you should decide what items you do not want/need to see, then direct how this paperwork should be routed in and out of your squadron. All paperwork is of some importance to your unit, but as a couple of commanders emphasized to me, some paperwork is more important than other!

Every commander has his or her own priorities regarding the importance of the many types of paperwork. A vast majority of commanders place performance reports (officer, enlisted, and civilian) on top of their lists. Nomination packages for awards and decorations usually come in second. Quality and timeliness are absolutely critical to both performance reports and nomination packages, and errors in either will get the paperwork quickly bounced back for correction. Worse yet, tardiness here can get you, the commander, in hot water and/or cause large problems for the ratee/nominee. I've seen a series of late reports be a significant factor in a commander's getting fired!

How to best ensure quality, timely packages? Many commanders recommend developing some type of internal system to accomplish and track these tasks.

A couple of examples as related to me: A flying squadron commander writing a performance report on one of his flight commanders started the process 6 to 8 weeks before the date the report had to be out of the unit. He first had a counseling/feedback session with the ratee and then asked the ratee to take a week and supply the commander with any additional information that might be applicable to the report. The commander told the ratee not to attempt to write the report himself.

Once the commander received the ratee's last input, he sat down and personally wrote the report. This particular commander was bound and determined to write these reports himself. He told me that on countless occasions during his career he had been told to write his own report and he wanted to eliminate that kind of action in his unit. After his secretary typed the first draft, the commander edited his own work and his secretary typed a second draft. This draft then went to the executive officer and the secretary for their edits. The commander's policy here was "no pride of authorship"; anything was fair game. His object was to obtain a good, solid critique and edit of an important document. The performance report bounced around the offices of the commander, executive officer, and secretary a couple of times until they were all satisfied with the product. The commander certainly was the driving force and had the last say, but he felt the involvement of his two trusted workers greatly added to the quality of the final product. The performance report normally left the squadron two to three days early. It was always hand-carried to the next location, never sent through distribution.

The commander in this example told me this system worked well for his two years in the saddle. Everyone had his or her job—the executive officer kept close track of upcoming reports and closely coordinated actions with the CBPO, the secretary kept precise files and computer disks on the reports, and the commander stayed on top of and ahead of his tasks. Teamwork paid off!

Second example—a decoration nomination package is being prepared on a master sergeant, a branch chief in a large maintenance squadron. The commander in this story told me his first action was to meet with the unit superintendent (a chief master sergeant) and the maintenance supervisor (a major) to decide what decoration was appropriate for the departing individual. This meeting occurred 6 to 8 weeks before the package was due out of the unit. The executive officer, as in the previous example, was responsible for tracking all suspenses of nomination packages. In this case, the decision was for the Meritorious Service Medal, which requires a good deal of paperwork and approval off base.

This particular squadron also had an awards and decorations NCO who was responsible for initially writing and working these packages. The commander would next sit down with this NCO to explain what type of nomination would be developed, any particulars the commander wanted incorporated into the package, and when the first draft was due to the secretary for typing. The commander also suggested who the awards and decorations NCO should see in the squadron to gather particulars for the nomination justification letter.

After the first draft was typed, the commander asked the superintendent and the supervisor to review the documents for substance and edits. Any changes here were typed into another draft that went to the awards and decorations NCO and then back to the commander. This commander related that having four to five people review the document during this process eliminated many typos and errors that one person reading and rereading a document will often miss.

While all this was going on, the executive officer was checking and double checking applicable dates and details for the necessary forms included in these types of packages. This commander also aimed to get the completed package out of his unit and up the chain of command a few days early if possible. He also told me that like the commander in the previous example, he never trusted the base distribution system with this paperwork—he always had it hand-carried to the next destination.

In both of the examples above, the commander developed some type of system to work the packages. Both commanders related they had some difficulty getting each system established; however, once the squadron had accomplished a few sets of paperwork, the task became much easier. One commander even said that after a few months the system ran on autopilot! Systems such as the two mentioned above may not be possible in every Air Force squadron. Not every unit has the luxury of an executive officer and a secretary; some units may only have one, others neither. Regardless, if possible, establishing some mechanism for handling performance reports and nomination packages should make life easier for the commander.

But what about the rest of the paper in your in-box? Opinions vary greatly here on what is important and maybe not so important. You're the commander, it's up to you to decide how to make it flow.

PROVERBS FOR COMMAND

"Administratively, don't hold things up, be decisive."

"You are the oil for the squadron machine."

One former commander reminded me of one important aspect of the paperwork flow—how you work your in-box affects others in your unit, especially your secretary and/or exec. A commander who sits on and procrastinates over paperwork can drive people nuts as they wait for some movement of documents out of the boss's office. On the other hand, a commander who automatically signs off on anything he reads without giving the content much thought can get severely burned some day!

I can offer no pieces of paperwork advice that are guaranteed to work in every situation. I will, however, offer a few TIPS/anecdotes relating to various types of paperwork you, as a commander, are likely to encounter in your in-box:

1. A request from your boss is important and normally goes to

the top of the pile. It's smart to work these fast and accurately; it's also not wise to leave the reply solely to someone else. Stay involved until you are certain your boss has what he needs. Each reply should leave the squadron under the commander's signature.

2. One commander had his administrative personnel keep a paperwork suspense log on the computer. Quick daily updates helped the commander stay on top of the situation.

3. A commander will often receive "reply by indorsement" (RBI) letters from all sorts of agencies. Many of these are routine and can be answered in seconds. Others will irritate you and waste your time. One commander got tired of receiving RBIs from staff sergeants telling him, a lieutenant colonel squadron commander, to reply by indorsement by such-and-such a date. After a couple months in command, he began to immediately trash any such correspondence. A couple of months after that, his unit was "highlighted" on a slide during the wing staff meeting for not replying to numerous requests. At this point the squadron commander explained the details and expressed his concern that such RBIs were fine for a large department store but not a military organization. From then on, RBIs ceased and correspondence was worked in a military manner!

4. Another type of paperwork you will likely encounter involves personnel in your squadron who are missing appointments. No commander enjoys getting these notices; they only create headaches and are likely to show up on someone's slide at another wing staff meeting. Before you jump down the alleged offender's throat however, fully check out the situation. More than one commander related that far more often than not, the correspondence was the result of a clerical mistake. One particular commander told how he kept receiving a rash of missed appointment letters from the dental clinic. After investigation, he learned that 95 percent of these letters were issued in error; either the individual had met the appointment or the appointment had been canceled weeks in advance. The commander called the dental clinic commander, explained the situation and asked for help. This cycle repeated itself for three more months until the squadron commander decided it was time to play hardball. At the next wing staff meeting, the wing commander sternly chastised the

squadrons for having so many missed dental appointments during the last quarter. This particular unit commander then produced his own set of slides depicting exactly how many of these appointments were actually missed and how many were the clinic's clerical foul-ups. The remainder of the day was not pleasant at the dental clinic! Bottom line—thoroughly check out these types of situations first; if your people are at fault, take action; if someone else is at fault, take action.

These are just a few examples of paperwork situations you may run into as a commander. Remember, you make the call on how to prioritize your in-box. Finally, TIP—don't let the paperwork tie you to your desk—your mission and people are far more important!

Writing inside the Squadron

While verbal communication can get a message to your people quickly, it sometimes lacks the preciseness of a well-written communication, especially when the topic is somewhat complicated. Also, written communications tend to be more permanent since they can be referred to as often as needed. For these reasons, squadron commanders have always used the written word to communicate with their unit on a wide variety of issues. Written communications in a squadron can take many forms: unit regulations/operating instructions, policy letters, newsletters, flyers, notebooks, and single notes.

Squadron regulations and operating instructions (OIs) are a more formal method of delineating standards and policies to your unit. Many bases and wings have set up elaborate series of such publications to help in the everyday conduct of business. Other units have no regulations or OIs. They can be very useful when kept up to date, standardized, and followed correctly. However, there is one aspect of these documents that every commander needs to consider: an inspection team can hold you responsible for following exactly what you have spelled out in any regulation or operating instruction. You are giving the IG more ammunition with which to inspect your operation!

I discarded all unit regs, replacing them with policy letters for which an IG team couldn't hold the squadron strictly accountable. I also kept these letters down to the minimum necessary to run the squadron effectively. I believe in giving people the freedom to make decisions on their own concerning unit operations and not tie them down with overly restrictive guidance.

A monthly newsletter can focus people's attention on particular topics. Colonel Lorenz had a very effective newsletter that he used to keep his people focused on their mission. He covered various topics in his monthly paper and always included articles on leadership and safety. Steve usually wrote a leadership article himself, and he always had his crews write a column on safety from personal experience relative to their KC-135 mission. He got a good portion of the squadron involved; therefore, the end product had meaning to the entire unit.

Col Bill Welser's C-141 unit had always used a lot of single-paged flyers placed in people's mailboxes to "get the word out." This works well if everyone regularly checks his mailbox, the distribution system works correctly, and people bother to read everything in the mailbox. Bill said that when he took over he saw that the flyers were often not read or were immediately trashed. To enhance written communication in the unit, he put up a few bulletin boards devoted specifically for information he needed to get out to everyone. These were not the usual boards with the base IG's picture on it or the board with the social actions phone number: those were elsewhere. Colonel Welser found that when he cut down on the overall number of flyers and posted them on these new boards, people started to read the information. There was also far less trash around the squadron.[10]

Lt Col Don LaMontagne used a similar technique in his F-111 unit. He set aside a special notebook that contained a record of every wing/base meeting that anyone from his unit attended, a record of all squadron meetings, any suspense for which his squadron was responsible, and other information he deemed necessary. His staff was responsible for signing off the notebook, but it was open to everyone. After a while, most everyone in the squadron was reading it on a regular basis.[11]

A final, more direct way for the commander to communicate with individuals or small groups is writing them a short note. Nothing formal here—the note doesn't have to be typed, just make sure the recipient can read it. I picked up an effective technique from Col Tom Whaylen when I was his operations officer. Every time he wrote anything in the squadron, no matter what the topic or who was to get the note, he always used a purple felt-tip pen. Then, whenever anybody got a note written in purple they knew it came from the commander. This practice proved effective and eliminated confusion. I tried the same thing when I took over using red and believe me, it works.

Remember the bottom line on unit communications—get the total, ungarbled word from the top to the bottom or vice-versa as quickly as possible.

Other Dealings with Outside Agencies and Units

After about one hour in command, you should realize that much of your work will be concerned with what I'd call lateral communications—communications with people or groups upon whom you depend for support to run your squadron. Every Air Force squadron is supported by and is in support of other units. No islands out there!

In dealing with these "lateral communications," the same guidance holds true for ensuring accurate and timely information flow inside a unit and up the chain of command.

Take, for example, your squadron's supply account. This is a type of communication (written documentation) that lists the Air Force property for which your unit is responsible. Normally, the commander does not personally handle this list on a day-to-day basis; however, the commander is responsible for its accuracy. To be responsible for its accuracy, it certainly helps to know a little something about the supply system and how your squadron interfaces with the base supply squadron. You, as the commander, certainly do not need to become an expert on the matter, but a short briefing from the supply squadron will not hurt and may save your backside come inspection time!

There are many examples of other areas where a commander needs a good overall understanding of the basics, but not all the complicated details—budget, finance, CBPO, family support, and so on. Taking the time early in a tour to become familiar with those topics will enable a new commander to work these lateral communications efficiently.

Notes

1. Lt Col Ray Chapman, interview with author, Ft McNair, D.C., 19 September 1989.

2. Lt Col Steve Lorenz, interview with author, Ft McNair, D.C., 16 October 1989.

3. Lt Col John Bell, interview with author, Ft McNair, D.C., 11 October 1989.

4. Lt Col Ron Love, interview with author, Ft McNair, D.C., 1 November 1989.

5. Lt Col Bob Lawrence, interview with author, Ft McNair, D.C., 9 November 1989.

6. Lt Col Dave Morrow, interview with author, Ft McNair, D.C., 21 September 1989.

7. Bell interview.

8. Col John Brooks, interview with author, (no additional info).

9. LTC Steve Rippe, interview with author, Ft McNair, D.C., 2 November 1989.

10. Col Bill Welser, interview with author, Pentagon, 27 October 1989.

11. Lt Col Don LaMontagne, interview with author, Ft McNair, D.C., 5 October 1989.

Chapter 5

Finishing the Job and Going Out in Style

One morning in mid-February 1989, I received a call from wing headquarters informing me the wing commander wanted to see me in an hour. I asked the nature of the meeting and his secretary said she didn't know. The wing commander's secretary *always* knew the nature of his business, so I figured something was up—like my body was about to be moved! I had been in command for almost two years, was headed to the National War College in the summer, and the wing was expecting a major inspection about the time I was due to depart.

Sure enough, the wing commander called me in and informed me the operations officer of another squadron would be taking command of my squadron in mid-March. He congratulated and thanked me for a successful command and told me I'd be moving up as deputy base commander for the short period until I left for War College. As I walked back into the squadron after that meeting, many thoughts and emotions raced through my mind, foremost being how to effect a smooth transition between myself and my successor.

Finishing the Job—Time Compression!

Regardless of how much actual time you have between the notification and your actual outgoing change of command, it never seems to be enough—you are constantly rushed! Time is always compressed at the end of a command tour because of the flood of events and requirements that engulfs the outgoing commander. Performance reports need to be closed out (officer, enlisted, and civilian), major squadron projects need to be completed, discipline problems should be brought to conclusion, your people should be thanked for their support—this list could go on and on.

After viewing the situation, the outgoing commander normally concludes there simply is not enough time to complete everything and it's best to organize and prioritize actions. One former commander explained how he broke these actions down into three categories.

Critical tasks—those items that need to be completed by the outgoing commander before the change of command; for one reason or another, the incoming commander cannot or should not accomplish these items.

Necessary tasks—items that the outgoing commander should complete before the changeover; the incoming commander could possibly accomplish these items, but the outgoing commander is much better suited to finish the task.

Nice-to-do tasks—items that the outgoing commander would like to accomplish before he leaves if possible; the new commander may also be able to accomplish these tasks easily.

The following are a few examples of various types of tasks that you, as an outgoing commander may be faced with before turning over the unit flag to a successor.

Critical tasks normally center around the squadron's mission and people. The first action is for *you* to tell your people of the upcoming change of command. I immediately called a special meeting of all available unit supervisors, told them the news, and asked them to spread the word quickly. I wanted my people to hear the news from within the squadron instead of through the rumor mill or from outside the squadron.

Another critical task is the completion of performance reports. Hopefully you've planned ahead and are not staring at a flood of reports—all being generated simultaneously because you are leaving. You will always have some reports to accomplish, and these reports deserve the same careful work that all others received; don't shortcut them because you may have a short suspense!

One commander who got caught on a very short notice assignment (to another good job) requested extra time to complete the necessary performance reports. He needed to be forceful with his arguments, and he finally convinced the bureaucracy, up through MAJCOM personnel, that he needed the extension to write a large number of important reports accurately and effectively.

Most commanders consider turning over a "fully up and running" unit to their successor a critical task. Lt Col Buzz Moseley told me his main concern at the end of his tenure was to turn over an F-15 unit that was "all together," functionally and administratively. He wanted all actions regarding their mission completed, to include the final revision on their new syllabus, which had been a major, ongoing project. Also, he aimed at completing all performance reports, decoration packages, personnel paperwork, and post inspection reports before the change of command. He also ensured that all supply and budget accounts were squared away. The object: to finish the job with a totally fixed, smoothly running squadron.

Four former commanders specifically emphasized clearing up all discipline and UCMJ matters before the change of command—they considered this task critical. Nobody wants to inherit a lot of problems, and the commander who initiated any discipline action is in by far the best position to complete the action in a consistent, just manner. One of these four commanders had himself inherited a serious personnel discipline problem when he took over his unit. He said he had difficulty completing the action effectively because the case involved a long history of circumstances that he could never fully ascertain. He was bound and determined not to hand over any such problems to his successor.

Finally, and most important, one commander said that during his last week, he went around the squadron, all shifts, day and night, thanking each and every member for their hard work and support. Thanking each member of your unit individually may be impossible and fall under the nice-to-do category, but, in some way, thanking your people for their support is critical.

A few items I'd normally classify as *necessary*.

With regard to the unit's mission itself, as commander, ensure any updates to unit procedures or policies are completed. If your operations officer, supervisor, or superintendent are not departing the squadron with you, the mission is usually not a problem. They will provide mission continuity during the transition of commanders, so the daily operation itself should just continue along unless there are unusual circumstances in progress, such as a major revision of the mission.

I talked with one training squadron commander who ran into this exact situation. His unit was changing five separate phases of its syllabus to include not only updates for new equipment but also a major shift in training philosophy. He explained how once a change of command date was set, he ensured that two of the five phases were fully completed by then, and left most of the other three phases for the new commander to work. He felt this was better than turning over five partially completed phases to his replacement.

People—if possible, at some point before the change of command, stop all personnel moves inside the squadron. The new commander deserves the right to make some of these decisions himself; after all, he'll be in charge for the next two years or so. One commander who was being replaced by his operations officer told me he let his successor make most of these decisions during the last six weeks before the changeover.

Any awards and decorations packages should be completed before the change of command. As outgoing commander, you are in a much better position to judge the accuracy of these packages than your successor. Also, if you are departing base, you may get involved in your own nomination package!

If it is at all possible, the outgoing and incoming commanders need to get together and discuss the "state of the unit"—a necessary task. Most outgoing commanders will initiate this process by calling the new commander and offering their congratulations and any desired assistance.

The relationship between a unit commander and his or her successor can vary greatly and is largely determined by the incumbent. As an outgoing commander, you will undoubtedly have at least a few mixed feelings for your successor—your replacement may be your closest friend in the whole world, but will be taking over something that has grown to be a part of you. The opposite situation may also arise—you may totally dislike this officer and he or she may dislike you. Whatever the situation, for your unit's sake, make the best of it.

Your unit's personnel may already have some opinion about your successor also. Whatever that opinion is, as outgoing commander, you should try to build up your successor in a positive vein in their eyes. I talked to a couple of commanders who said this wasn't easy, but they gave it their best shot. TIP—remember back to when you were in your successor's shoes—make it as easy as possible for your replacement. If circumstances permit, meet as often as necessary to pass along lots of information about all aspects of the unit. Remind the new commander that your words reflect your opinions. If your successor is new to the base, share your thoughts on the "lay of the land"—the chain of command above, the other units on the base, and especially anything that might jump out and bite early.

If you are staying on base in another capacity, offer your counsel anytime the new commander has a question, but give reassurances that you will mind your own business after the change of command. It's also a good idea for your spouses to get together, if applicable.

Remember you're still the commander until you pass the flag. A couple of former commanders related how they had to "hold back" their replacements who were a little too anxious to get in the middle of squadron business before officially taking over. A few commanders told me they invited the next commander to the squadron for discussions only after duty hours so as to avoid placing their people into an awkward position with regard to loyalty. Other commanders related they personally escorted their successor around their squadron, introducing him to all members. Safe advice here is simply to handle the matter in whatever way is natural for you and the squadron.

Miscellaneous paperwork—the recommendation here is to complete as many paperwork actions as you possibly can. Most will fall into the "necessary" category, but you should decide what is "critical, necessary, and nice to do." Numerous commanders said they worked hard to leave an empty in-box!

There are many items that can fall into the nice-to-do category. It would be nice to have a final commander's call, a finis-flight, a final TDY or deployment with the unit, a final dinner with your unit supervisors, and so on. As outgoing commander, you know best how much time you have and how your priorities rank up—your decisions all the way.

Going Out in Style

The squadron will most likely want to throw you a farewell party, and their spouses may want to do something for your spouse also. Anything goes here. Your, and their, call—the only suggestion I offer is to have the function(s) before the change of command, not after, when you are no longer the commander.

Normally, the outgoing commander and squadron plan the change-of-command ceremony. Each base usually has a fairly set way of holding these affairs. Certainly follow the guidelines given to you by the wing commander, but go ahead and add your personal touch to the affair, something different—go out in style! One commander had his changeover outside by the base's new static aircraft display and the reception in the newly remodeled base museum, a first for both ideas.

Planning—one commander with whom I discussed this subject really emphasized nailing down the many details early. This may require a project officer/NCO. TIP—pick two, someone with experience and someone who deserves some limelight but maybe doesn't have the experience. Challenge them to make the ceremony one the unit will be proud of and one the wing will remember.

In the days following the ceremony try to stay away from the squadron—give the new commander room to operate. Take some leave if possible. Some of the commanders I interviewed told me they have never gone back to their old units, others related they waited a couple of months, while still a few others told me they had to go back on business fairly soon after the ceremony, but made their stay as short as possible.

One final topic—your speech. Usually after the actual change of command, the outgoing and incoming commanders are asked to say a few words to the audience. If your emotions have not gotten the best of you yet, they've got one more chance, and the odds are that they'll get you now. Most former commanders recommend making this speech short and sweet, concentrating on thanking the many people who supported you during your tenure as squadron commander.

I can still clearly remember the scene. As I began publicly thanking people, I looked out in the audience and saw my wife and two daughters, seated directly in front of the two men who had been my operations officers and their wives. Seven people who had spent many of their waking hours in the last two years supporting me. At this point I stepped back from the podium for what seemed an eternity until my eyes became dry again before I went ahead and finished my remarks.

I spent four years and ten months at Mather AFB, California, had many good days and some bad days; but 15 March 1989 will always remain my least favorite day. The best job I'll ever have in the United States Air Force was behind me.

Index

U.S. Government Printing Office: 2001— 633-007/22124